Plants, People, and Culture

PLANTS, PEOPLE, AND CULTURE

The Science of Ethnobotany

Michael J. Balick

Paul Alan Cox

**SCIENTIFIC
AMERICAN
LIBRARY**

A division of HPHLP
New York

Cover image: A detail from *History of Medicine* by Diego Rivera, a mural at the Hospital de la Raza, Mexico City, Mexico. Schalkwijk/Art Resource, NY.

Library of Congress Cataloging-in-Publication Data

Balick, Michael J., 1952–
   Plants, people, and culture : the science of ethnobotany / Michael J. Balick, Paul A. Cox.
      p.   cm.
   Includes bibliographical references and index.
   ISBN 0-7167-5061-9
   ISBN 0-7167-6027-4 (pbk.)
   1. Ethnobotany.   I. Cox, Paul Alan.   II. Title.
GN476.73.B35   1996
581.6 — dc20   96-6493
                                                                                                          CIP

ISSN 1040-3213

Printed in the United States of America

Scientific American Library
A division of HPHLP
New York

Distributed by W. H. Freeman and Company
41 Madison Avenue, New York, NY 10010
Houndmills, Basingstoke RG21 6XS, England

Second printing, 1999

This book is number 60 of a series.

ANTH
591

Contents

Preface

In this book, we assert that the very course of human culture has been deeply influenced by plants, particularly plants that have been used by indigenous peoples around the world. Although most schoolchildren learn, for example, that Columbus ventured out in search of a new route to the spice islands, few of us realize the tremendous geopolitical influence of the demand for spices on the rise and decline of the great cities of Europe. Many pharmacists know that plants once played an important role in healing, but few are aware that potent anesthetics and glaucoma treatments were derived from studies of arrow poisons and witch ordeals, respectively. Whether it be the construction of vast sailing boats used to transport Polynesians to new islands, or the role agricultural surpluses and scarcities played in the stratification of early societies, or the formulation of potent hallucinogenic snuffs used to transport Amazonian shamans to the other world, plants have largely guided the trajectory of human culture.

The study of the interactions of plants and people, including the influence of plants on human culture, is the focus of the interdisciplinary field of ethnobotany. The interests of ethnobotanists range from the functioning of indigenous healing systems to the imbibing of plants in rituals, from the cultural consequences of the extinction of a vine used to construct fish traps to the health consequences of a change in diet, from the class implications of forms of dress to the cultural role of body adornments. As scientists have become aware of the rich variety of questions that ethnobotany addresses, the field has been undergoing a resurgence. Now equipped with new scientific tools from molecular biology, mechanical engineering, and medical anthropology, modern ethnobotanists are asking a dazzling array of new questions while shedding new insight on older questions. Using the latest molecular techniques, for example, some ethnobotanists are empirically testing theories of plant origin, while others study the plants used by indigenous peoples in healing for clues to biochemical function, in the hope of developing better pharmaceuticals.

We write from the perspective of scientists who have spent a large part of our careers living in remote villages interviewing healers, weavers, shipwrights, and other indigenous experts in the use of plants. Over the last two decades, both of us have undertaken extensive fieldwork in the tropics, with Michael Balick focusing on Central and South America and the Caribbean, and Paul Cox concentrating on Oceania and Southeast Asia. Although much fine ethnobotanical work is being done in the deserts, temperate forests, and arctic regions of the planet, because of our backgrounds we have focused on the areas we know best, as will become particularly apparent in the last chapter, where we discuss indigenous strategies for tropical rain forest conservation.

We saw this opportunity to write for the Scientific American Library as a chance to convey to a general audience some of the profound insights ethnobotany offers into the human condition, as well as a chance to discuss some of the exciting new advances in this discipline. We also hope to share the joy—and occasional adventure—inherent in living and studying with indigenous peoples. As is customary for books in this series, footnotes and literature citations are not used in the text, in order to make it more accessible to the nonspecialist. Nevertheless, we wish to acknowledge the tremendous intellectual debt we owe both to our teachers and to our colleagues, most of whom we could not mention in these pages. Interested readers can introduce themselves to this broader community of ethnobotanists by reading some of the books and articles mentioned at the conclusion of this book, or browsing the pages of *Economic Botany,* the *Journal of Ethnopharmacology,* the *Journal of Ethnobiology,* or any of a number of anthropological, botanical, and chemical journals. We do, however, specifically wish to thank Anthony Anderson, Rosita Arvigo, Herbert Baker, Shayne Baker, William Balée, Sandra Banack, Bradley Bennett, Mark Blumenthal, Silviano Camberos S., Tom Carlson, Wade Davis, Elaine Elisabetsky, Thomas Elmqvist, Memory Elvin-Lewis, Nina Etkin, Richard Feinberg, Jay Feldstein, David Harris, Maurice Iwu, Joel Janetski, Timothy Johns, Stephen King, Walter Lewis, Will McClatchey, Dennis McKenna, Robert Mendelsohn, Gary Nabhan, Francoise Pierrot, Sir Ghillean Prance, Gerald Reaven, Beate Reuter, Laurent Rivier, Gunnar Samuelsson, Richard Evans Schultes, Gregory Shropshire, Bruce Smith, Susan Wilcox, Sheryl Wilson, and David Wright for useful criticism, assistance, and encouragement. Our research was assisted at Brigham Young University by Marilyn Asay, Rebecah Davis, Andrea Dewey, Kim Hart, Amy Hettinger, Richard Jensen, Alexandra Paul, and Mark Philbrick; and at The New York Botanical Garden by the scientific staff, particularly Daniel Atha, Hans Beck, Brian Boom, Douglas Daly, Roy Halling, Patricia Holmgren, John Mickel, Scott Mori, Christine Padoch, Chuck Peters, Jan Wassmer Stevenson, and Barbara Thiers, as well as Willa Capraro, Mee Young Choi, Sandi Frank, Susan Fraser, Cori Morenberg, John Reed, and Muriel Weinerman. We are particularly grateful to the editorial and production team at W. H. Freeman, including Travis Amos, Robert Biewen, Jonathan Cobb, Julia DeRosa, Christine

Hastings, Judy Levin, Susan Moran, Bill Page, Mary Shuford, Allyson Siegel, and Vicki Tomaselli, who, for a moment in time, allowed us to turn their headquarters at 41 Madison Avenue into a *hui,* a traditional meeting in which stories and legends of plants and people are recounted. We are grateful to our families for their encouragement and willingness to give us time off to work on this book. We particularly thank the Montgomery Foundation, for providing us with a peaceful setting in which to begin this book, and Francesca and Bradley Anderson, who offered their home to Michael Balick while he worked on the manuscript.

We thank those institutions that have supported our work for so many years. In particular, Paul Cox wishes to acknowledge Daniel and Kathy Betham, Brigham Young University, the Danforth Foundation, the Institute for Polynesian Studies, the Miller Institute for Basic Research in Science, the National Cancer Institute, the National Science Foundation, the Schering Research Institute, the University of Melbourne, Verne and Marion Read, the University of Umeå, and the University of Uppsala for their generous support of ethnobotanical research. Michael Balick wishes to thank the National Cancer Institute, the U.S. Agency for International Development, the Metropolitan Life Insurance Foundation, the Overbrook Foundation, the Edward John Noble Foundation, the Rex Foundation, the Rockefeller Foundation, Susannah Schroll, the John D. and Catherine T. MacArthur Foundation, the Gildea Foundation, the Nathan Cummings Foundation, the Charles A. and Anne Morrow Lindbergh Foundation, and the Philecology Trust, through the establishment of the Philecology Curatorship at The New York Botanical Garden.

To verify the plant taxonomies in this volume we have relied for species names on *Index Kewensis,* for genus names on D. J. Mabberley's *The Plant Book,* and for family names on the late Arthur Cronquist's system. All names have been checked against the Gray Herbarium Card Index and *Index Nominum Genericorum.* Titles and spellings of pre-Linnean works have been checked against the *Catalogue of the Library of the British Museum of Natural History.* Family designations are shown in the text in brackets when a taxon is mentioned for the first time.

Finally, we wish particularly to thank those indigenous people who for years have so unselfishly taught us about plants, people, and remaining both human and humane in a complex world. We dedicate this book to our friend and former classmate, the late Calvin R. Sperling, plant explorer extraordinaire.

Michael J. Balick
Paul Alan Cox
March 1996

Diego Rivera depicted ancient Aztec healers and some of the medicinal plants they used in this detail from his mural *History of Medicine,* found in the Hospital de la Raza, Mexico City.

People and Plants

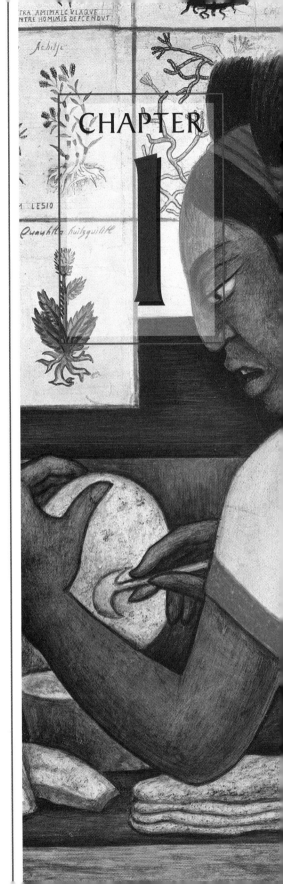

Within sight of Mount Everest, between the Ganges River and the foothills of the Himalayas, grows a small climbing shrub with pinkish-white flowers, smooth leaves, and milky sap. Called in Hindi *chotachand*, the shrub is rarely disturbed by the local people unless someone is bitten by a snake. The shrub is then unearthed and its long root given to the victim. A local legend claims that in ancient times mongooses were observed to feed on the plant before engaging in combat with cobras. Copying the reputed activity of the mongoose, local people found that the shrub could serve as a potent antidote to snakebite.

Eighteenth-century botanists named the shrub *Rauvolfia serpentina* [Apocynaceae]. *Rauvolfia,* the name of the genus, honors the sixteenth-century physician and botanist Leonhard Rauwolf; *serpentina,* the name of the species, describes the snakelike appearance of its root. Apocynaceae is the name of the family of plants that includes *Rauvolfia serpentina.*

Rauvolfia serpentina, the snakeroot plant traditionally used as a sedative in the Ayurvedic medicine of India.

In the eighteenth century, a specimen was sent to the herbarium, a museum-like repository of dried plants in Leyden, Holland, but *R. serpentina* was then ignored by scientists, as most medicinal plants have been: less than ½ of 1 percent of all flowering plant species in the world have ever been exhaustively studied for their potential pharmacological activity.

But the shrub was not forgotten by the local people, and its use spread to nearby cultures. In Bihar province it was rumored that when a demented man had eaten slices of the root, he was cured of his madness. The people of Bihar began to use the plant to treat insanity, epilepsy, and insomnia, calling the shrub *pagal-ka-dawa,* "insanity cure." They also found that a single dose of powdered roots could put a child into a deep sleep that lasted all night. Slowly the use of *R. serpentina* to treat anxiety, insomnia, and madness spread throughout India.

In 1931, Indian chemists isolated a variety of molecules from the plant, but found them to be relatively inactive. Their interest was renewed, however, by a report published in the *Indian Medical Record* that *Rauvolfia* powder not only had a hypnotic effect but also dramatically lowered blood pressure. Like many scientific reports published in developing countries, this discovery was ignored by Western scientists. Then in 1949 Emil Schlittler, a chemist at CIBA pharmaceuticals in Basel, Switzerland, read a clinical study of *Rauvolfia* by R. J. Vakil in the *British Heart Journal.* Schlittler, with his colleague Hans Schwarz, extracted from *Rauvolfia* roots an alkaloid, a nitrogen-containing physiologically active organic compound, which they named reserpine. They demonstrated that at remarkably low oral doses, 0.1 milligram per kilogram body weight, reserpine lowered blood pressure. In clinical tests reserpine lowered one patient's blood pressure from 300/150 to 160/100. American investigators confirmed these dramatic findings. "It has a type of sedative action that we have not observed before," a Boston team reported to the New England Cardiovascular Society. "Unlike barbiturates or other standard sedatives, it does not produce grogginess, stupor, or lack of coordination."

CIBA soon introduced reserpine to medicine under the trade name Serpasil. Up to that time, all known compounds that lowered blood pressure did so by dilating blood vessels. Reserpine, however, had a direct effect on the hypothalamus of the brain, opening up an entirely new mode of pharmacological action. In 1954 the New York Academy of Sciences sponsored a symposium devoted to the pharmaceutical importance of *Rauvolfia.* Reserpine became the first major drug to treat one of the most serious illnesses of the Western world: hypertension. More recently it has been prescribed in combination with other antihypertensive drugs, such as hydralazine hydrochloride.

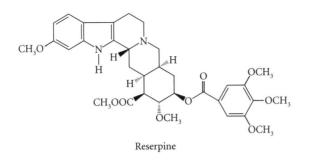

Reserpine

The alkaloid reserpine, a chemical derived from *Rauvolfia serpentina,* has become a major drug for treating high blood pressure.

How are we to characterize the discovery of reserpine? Does discovery of this important drug rest on "solid" science, such as structural chemistry and pharmacology, or is it attributable to folklore and legend? Laboratory scientists may hail the invention of reserpine as serendipitous, but one fact is inescapable: a plant used by indigenous peoples eventually became the source of one of the world's most important pharmaceuticals.

There appears to be a wide gulf between folk knowledge and modern science, a gulf based on empirical verification. Science is the acquisition of knowledge based on careful observation and experimental tests of theory. Indigenous traditions are sometimes derided as steeped in superstition. Nevertheless, every time a Shipibo hunter fires a poison dart at an animal or a Tahitian healer administers a medicinal plant to a sick child, the efficacy of the indigenous tradition is empirically tested. It appears that indigenous traditions and science are epistemologically closer to each other than Westerners might assume. The contexts of trials performed by Western scientists and by Shipibo Indians or Tahitian healers are obviously very different, but the empiricism in both is of interest. The field of study that analyzes the results of indigenous manipulations of plant materials together with the cultural context in which the plants are used is called ethnobotany.

In broad terms, ethnobotany is the study of the relationships between plants and people. The two major parts of ethnobotany are encapsulated in the word itself: "ethno," the study of people, and "botany," the study of plants. Arrayed between these two points labeled "ethno" and "botany" lies a spectrum of interests ranging from archaeological investigations of ancient civilizations to the bioengineering of new crops. However, the field is limited on both sides. On the botanical side of the field, few ethnobotanical studies are concerned with plants that have no connection to people. On the ethno side, most studies are concerned with the ways indigenous people use and view plants. And those uses and those views can provide deep insights into the human condition.

The American botanist John W. Harshberger coined the term "ethnobotany" in 1895 to describe studies of "plants used by primitive and aboriginal people." His 1896 publication, *The Purposes of Ethnobotany,* is generally accepted as a starting point for this field as an academic discipline.

The Samoan healer Fa'ifili Fagomano, of Tafua village, demonstrates the use of *Cordyline terminalis* [Agavaceae] to treat a fever in a sick child.

Much of ethnobotany deals with intellectual goals similar to those of cultural anthropology: to understand how other peoples view the world and their relation to it. The way people incorporate plants into their cultural traditions, religions, and even cosmologies reveals much about the people themselves. Some products of ethnobotanical research, such as reserpine, transcend mere anthropological interest and are of profound importance to the West. People use plants in so many different ways that there are few arenas of human endeavor in which plants do not play an important role. Indeed, plants have determined the very course of civilization. In the thirteenth century, Marco Polo described an island "producing pepper, nutmegs, spikenard, galingale, cubebs, and all the precious spices that can be found in the world." This report spurred a search for the Spice Islands, which inadvertently resulted in Europeans' discovery of America and culminated in Magellan's circumnavigation of the globe. Since the Renaissance, patterns of international trade in rubber, opium, and quinine have altered the fates of entire nations.

Even the plague of drug abuse that afflicts Western countries today can be considered an ethnobotanical problem, since it involves illicit traffic in substances—heroin, cocaine, hashish—that are derived from plants that indigenous peoples have used for centuries. In the economic sphere, few industrial societies can ignore the pivotal role of agriculture and forestry; indeed, much of the new regulatory structure of the European Union governs trade in crops and other plants. A plethora of environmental crises—global warming, loss of biodiversity, tropical deforestation—are, at their core, issues involving plants. Ethnobotanical research may shed light on some of these issues, and may even point to possible solutions. Although we hope to make a convincing case for the importance of plants in the development of civilization, we will focus most of our attention on ethnobotanical studies of indigenous peoples.

The term "indigenous peoples" refers to peoples who follow traditional, nonindustrial lifestyles in areas that they have occupied for generations. Thus the European settlers of Australia or North America are not considered "indigenous," while the Australian Aborigines and the American Indians are. Given the pivotal role of plants in directing the trajectory of Western societies and the ubiquity of Western cultures today, why should ethnobotanists focus so much attention on indigenous peoples?

There are several reasons for this interest. First, the relationships between plants and people are often clearer in indigenous societies than in our own, since the link between production and consumption is more direct. Within a single village an ethnobotanist can study how people forage for wild plants or sow crops, how they use plants to construct houses, baskets, boats, or clothing, and the role plants play in myth and lore. In these cultures, such information resides within individuals, families, or villages. In industrialized societies, however, economic patterns of production and consumption are so complex that most individuals have little understanding of the botanical origins of or pro-

Julie Chinnock, an ethnobotanist working with a Kekchi Maya shaman in the Punta Gorda region of Belize, prepares plants used to treat diabetes. The shaman had described the preparation and use of the plants to a medical doctor working as part of a team with Chinnock.

cessing technology used to provide the plant materials they use every day. Such information does not reside within extended families or even entire towns.

Take such a simple and ubiquitous object as a pencil. Not only has the philosopher Leonard Read failed to find any single individual in America who could accurately describe how to make a pencil, he discovered that companies that manufacture specific components of pencils are ignorant of how other components are produced and how they are fitted together to form an entire pencil. Such compartmentalization of knowledge is rare in indigenous cultures: though individual villagers may lack esoteric knowledge of a plant, they can usually refer an investigator to a local expert—a shaman, or shipwright, or weaver—who has the requisite understanding.

Second, indigenous cultures sometimes represent living analogues of the prehistorical stages of Western civilizations. Thus archaeologists can explore hypotheses concerning the hunter-gatherer phase of the earliest Europeans by studying the lifestyles of modern hunter-gatherers. They can scrutinize the precursors of modern agriculture by studying the pre-agricultural strategies of indigenous cultures such as that of the Torres Straits Islands, south of New Guinea. The problem, of course, is that we can never be sure how close (or distant) such analogies are. At least, however, they can generate useful discussion and help us reject unworkable hypotheses.

Third, indigenous cultures retain much knowledge concerning plants that Western peoples have largely lost. Indigenous peoples have of necessity maintained knowledge of plant medicines, textiles, and plant cultivation strategies. Some knowledge, such as ethnotaxonomic systems (biological classification schemes used by indigenous peoples) or legends and myths concerning plant origins, are of interest because of the insight they shed on the cultures themselves. Other types of knowledge can be of more immediate benefit to Western peoples. Indigenous knowledge systems, for example, can guide the development of new crop varieties or medicines.

Fourth, indigenous peoples are stewards of some of the most sensitive ecosystems on this planet. Indigenous knowledge systems, developed over centuries of residence in such habitats, can inform current debate concerning the conservation of natural resources.

Finally, in today's global economy, indigenous peoples are vulnerable to rapid economic and cultural change. Understanding of traditional ways, including uses of plants, can point to strategies for ameliorating negative consequences of that change.

Given this twin focus on plants and indigenous peoples, the ideal ethnobotanist is a combination anthropologist, archaeologist, botanist, chemist, psychologist, ecologist, explorer, folklorist, pharmacologist, and diplomat. Only through an interdisciplinary approach can we hope to understand the close connection between plants and human societies.

Plants as the Material Basis for Human Culture

One may ask why plants rather than animals have traditionally been the focus of such investigations. Why does ethnobotany command far more academic interest than its sister discipline of ethnozoology?

The material culture of nearly every people on this planet is based more on plants than on animals. From the Vikings of Scandinavia with their large wooden sailing vessels to the Maoris of New Zealand with their intricately carved meetinghouses, from the Shipibo Indians of the Amazon rain forest with their 3-meter-long blowguns to the Navajo of the North American desert with the dyes that color their patterned rugs, the peoples of the earth have long depended on plants for food, clothing, shelter, transportation, medicine, and ritual. Why should plants, rather than animals, play such a crucial role in the development of human cultures?

In many parts of the world, mats woven from plants such as *Pandanus tectorius* [Pandanaceae] serve not only as a floor covering but for sleeping, sitting, ceremonial exchange, and even boat sails.

Part of the reason is found in the profound ecological disparity between plants and animals: plants are able to transform atmospheric gases and minute quantities of inorganic nutrients into life itself. As a result, the plants outweigh all the elephants, lions, squirrels, and every other form of animal life by a factor of at least 10. Plants are also vast factories of chemical diversity. But the crucial difference is that plants produce while animals consume.

All animals, including people, depend on consumption not only for their lives but for the way they live. It is the food animals eat, be it herb, fowl, fish, or other flesh, that largely determines their position in the ecological community. In this ecological sense, animals are indeed what they eat and are defined in the web of life by what they consume. With the exception of a few microorganisms that contain chlorophyll and a few coral reef organisms that harbor symbiotic algae, no animal can live and grow without consuming something else. If we exclude carnivorous plants such as the Venus's-flytrap, and plants that live as parasites on a host, such as the mistletoes, we can distinguish plants not by what they consume but by what they produce.

But what about soil and nutrients? Is earth itself the food of plants? In 1648 the Flemish plant physiologist Johannes Baptista van Helmont elegantly tested this conjecture. Van Helmont planted a 5-pound willow in 200 pounds of soil. After five years he removed the willow and found that it weighed 169 pounds 3 ounces. Reweighing the soil after it had dried, he found that it now weighed 199 pounds 14 ounces. In other words, 164 pounds 3 ounces of willow was produced by only 2 ounces of soil. How could this be? Did the plant create its bulk from thin air?

The answer, of course, is yes. During photosynthesis plants are able to recombine the carbon atoms from carbon dioxide into the six-carbon rings we know as hexose sugars. Other elements used in the light and dark reactions of photosynthesis are important, but they are required only in minuscule quantities. The hexose sugars produced by photosynthesis are linked to make long-chain polymers such as starch and cellulose. When cellulose is combined with plant resins, it forms one of the most important building materials ever discovered: wood. The photosynthetic pathway leading from carbon dioxide to wood means that even the massive rain forests of the Amazon have ultimately been produced from thin air.

The carbon dioxide in the atmosphere is nearly inexhaustible. Thus plants compete with one another not for this gas but for a position in the plant canopy that will enable them to capture the sunlight that powers the photosynthetic process. Since they must maintain both their place in the sun and their roots in the soil, no terrestrial plants are mobile. Their immobility coupled with their tremendous production of cellulose makes plants a far more efficient and reliable source of building materials and food than animals. Carpenters need not chase their trees through the forest. Pity the culture that would require wildebeest bone or tiger skin for housing material or clothing. In times of abundance such materials might be found, but in times of scarcity, plants are far more accessible. As omnivores we can eat meat, and indeed, most cultures revel in the hunter's exploits or the fisherman's success. Yet all cultures, except for a few confined to tundra regions of the Arctic and pastoral peoples such as the Masai of Kenya, depend on plants for the bulk of their diet. Even peoples that follow their herds across the landscape depend on plants for forage for their animals.

People rely on plants for much more than food and shelter. The 250,000 species of flowering plants differ not only in form but also in hidden biochemistry. No animals, including human beings in white laboratory coats, have ever

Antonio Cuc, a *yerbatero* or herb gatherer in San Antonio Village in Belize, chops roots from *Chiococca alba* for use that day by a traditional healer. *Chiococca alba*, [Rubiaceae], called skunkroot by the villagers, is a powerful plant with many uses in the region.

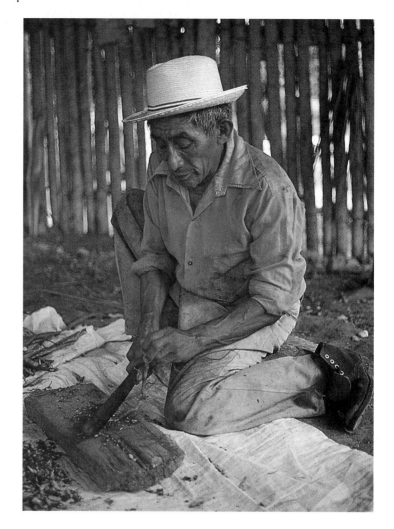

been able to produce even a fraction of the dazzling array of molecules routinely assembled by plants. Even today we cannot synthesize many of these natural plant chemicals. Since all plants require the same staples—carbon dioxide and sunlight—the biochemical diversity of plants probably has little to do with the machinery of photosynthesis. Among all plant species, only a handful of photosynthetic pigments have been discovered. Why, then, should plants produce exotic chemicals ranging from opiates that deaden our nervous systems to sweetening agents that enhance our diets?

Perhaps the answer again lies in the immobility of plants. Unlike animals, plants cannot move about to carry out reproductive activities or escape enemies. Plants must either rely on the uncertain forces of wind and water to move their pollen and seeds or alternatively entice animals to perform these services for them. In this sense, then, the flower of the orchid and the fruit of the mango tree represent contracts between plants and animals. The orchid provides sweet nectar and occasionally sex pheromones or a trysting place for insects that transport pollen. The mango provides needed nourishment to the flying fox that carries its massive seed.

Not all plant-animal interactions, however, are benign. The animal mouth, be it insect mandible or mammalian jaw, represents a potent threat to a carbohydrate-rich, water-filled organism that cannot flee a potential predator or parasite. Plants have therefore of necessity become specialists in animal biochemistry. Their chemicals function not only to reward animal pollinators and carriers of their seeds but also to repel, maim, or poison those animals that attempt to destroy them. The chemical agents that plants employ against animals have profound implications for medicine: we depend on these chemicals for 25 percent of our prescription drugs and for nearly all of our recreational chemical substances, including the caffeine in coffee, the nicotine in tobacco, the theophylline in tea, the theobromine in chocolate, and a virtual cornucopia of other psychoactive substances throughout the world.

This triad of immobility, carbohydrate production, and diverse biochemistry makes plants far more useful to human beings than animals can be. Indigenous peoples throughout the world have become expert at using the plant resources around them.

A display of traditional remedies, both plant and animal, from a market in Cuzco, Peru. Markets are fascinating places to carry out ethnobotanical investigations, and herb vendors are often quite knowledgeable about the origins of their materials and their various uses.

Plants and People in Ancient Times

Thatch for huts, timbers for boats, fibers for cordage, and textiles and dyes to color them all appear at early stages of human prehistory. Yet these uses pale in comparison with the use of plants for medicine and food. Agriculture is a relatively recent development in human history, arising independently in several parts of the world during the last 10,000 years. In valleys nestled in the Swiss Alps, groups of people lived in fishing communities on the edges of large lakes. Perhaps seeds from plants used for thatch or food fell onto piles of waste and flourished in the nutrient-rich environment. The ease of gathering these plants led people to repeat deliberately what began as an accidental process. The ancestors of the Polynesians in Indomalaysia developed a different form of agriculture by tending trees and tuberous plants that furnished edible nuts and rhizomes. Soon they discovered how to propagate these plants from cuttings, possibly by observing that the leafy tops of tubers grew after they had been discarded.

Cultivation of the grains that supported Western civilization also developed within the last 10,000 years. In what is now Iraq, remains of grinding stones from 8000 B.C. and seeds of wheat and barley from 6700 B.C. have been found. Mesoamerican agriculture also developed in the last 10,000 years: gourds and squash from 6000 B.C. have been found in Mexico; maize, one of the most productive plants in the world, developed even later. Yet the ancient use of such plants can be traced through more than archaeological conjecture: many societies have left graphical or written representations of their interactions with plants.

For centuries people created durable representations of plants, etching them in stone or molding them in clay. Such images not only provide modern ethnobotanists with clues concerning plant origins but function as tangible indicators of the importance these peoples attached to plants. Early beliefs about plants, such as the association of the creation of the first man and woman with

This rendering depicts the bas-relief at the ancient Palace of Nimrod, on the east bank of the Tigris River, portraying a winged god pollinating date palms.

a garden, were transmitted not only orally but also in written form. Such records often combine the religious and the pragmatic. An Assyrian bas-relief sculpture at the palace of Nimrod at Assur-Nassur-Aphi, for example, portrays winged gods pollinating date palms, evidence of early knowledge of plant reproduction. The sculpture also indicates, however, that the Assyrians viewed the process of pollination as part of the realm of religious experience. In the fifth century B.C. Herodotus recorded the Babylonians' knowledge of crop pollination with no religious overtones. Later Aristotle (384–322 B.C.) began philosophical consideration of plants. His student Theophrastus, who inherited Aristotle's library, wrote extensively about plants and recorded many observations made by a fellow student, Alexander the Great. The ninth book of his *Enquiry into Plants* contains a good deal of information about medicinal plants, although Theophrastus ridiculed the superstitions associated with the way the plants were gathered. Further efforts to codify mythical and folk medicine led Pedanius Dioscorides, a Greek physician of the first century, to write a compendium called *De Materia Medica,* which not only describes 500 medicinal plants but illustrates many of them. *De Materia Medica* was accepted as authoritative until the early Renaissance. Such early compilations of folk wisdom concerning medicinal plants were not confined to the West; traditionally dated as before 2000 B.C. the Chinese emperor Shen Nung compiled the *Pen Tsao,* which is perhaps the earliest known herbal.

Herbals and Medicinal Plants

In the early Renaissance there was an explosion of interest in herbals, most of which were based on the work of Dioscorides with incremental improvements made from the authors' own knowledge. The first herbal written in the Anglo-Saxon world was an eleventh-century codex known as the *Herbarium of Apuleius Platonicus.* The earliest printed English herbal was an anonymous quarto of 1525, printed by Richard Banckes: "Her beynneth a newe matter, the whiche sheweth and treateth of y vertues and proprytes of herbes, the whiche is called a Herball."

A year later a translation of a French herbal was published by Peter Treversi, and in 1538 William Turner published *Libellus de re Herbaria Novus.* In 1551 Henry F. Lyte (author of the hymn "Abide with Me") published a translation of Rembert Dodoens's herbal *Stirpium Historiae Pemptades Sex,* which had achieved renown on the continent because of its encyclopedic scope and superb plates of flowers. But the most popular of all sixteenth-century herbals was that of John Gerard, published in 1597. It is one of the few books to remain in print for over 400 years and is one of the most important books on plants ever published in the English language.

Gerard was born in Nantwich, Cheshire, in 1545. At the age of 16 he began a seven-year apprenticeship to a London barber and surgeon, Alexander Mason. Gerard traveled briefly as ship's surgeon in the Baltic but showed far more interest in plants than in life at sea. In 1577 he was appointed superintendent of the gardens of Lord Burleigh at the Strand in London, and later he was appointed curator of the Physic Garden of the College of Physicians of London. George Baker, surgeon to Queen Elizabeth, reported that Gerard's own garden was filled with

> all manners of strange trees, herbes, rootes, plants, flowers, and other such rare things, that it would make a man woonder, how one of his degree, not having the purse of a number, could ever accomplish the same. I protest upon my conscience, I do not think for the knowledge of plants that he is inferior to any.

In 1596 Gerard published a catalogue of the plants in his garden. The work he published the next year, *The Herball, or Generall Historie of Plantes,* rapidly became one of the most quoted botanical works ever published. Gerard's *Herball,* with 1392 pages and 2200 woodcut images of medicinal plants, was greeted with tremendous enthusiasm by medical practitioners, who prescribed from it.

Twentieth-century authors have accused Gerard of plagiarism because much of his herbal appears to be taken from the earlier work by Dodoens. This charge is disingenuous, however, for herbals are by definition compilations of knowledge accumulated through the ages. Dodoens himself borrowed very heavily from both Pliny and Dioscorides, who in turn borrowed from the *rhizotomi,* Greek root gatherers, whose business was "preparing and selling of roots and herbs that were of repute in medicine."

It was precisely this collection of accumulated folk knowledge that made John Gerard's *Herball* so valuable. Renaissance doctors carefully searched its pages for descriptions of plant medicines. Take, for example, the entry on page 646 for the foxglove plant, *Digitalis purpurea* [Scrophulariaceae]:

> Foxe gloue boiled in water or wine, and drunken, doth cut and consume the thicke toughness of grosse and slimie flegme and naughtie humours; it openeth also the stopping of the liver, spleene and milt and of other inward parts.

The frontispiece of Gerard's *Herball.* First published in 1597, this compilation of information on medicinal plants was extensively referred to by physicians in search of herbal remedies.

William Withering and Cardiac Drugs

Gerard's claim for the effects of foxglove on internal organs was not examined until 1775, nearly two hundred years later. After a decade of investigation, William Withering published in 1785 *An Account of the Foxglove and Some of Its Medical Uses.* Withering quoted Gerard's account of the "vertues" of foxglove and proposed that the plant could yield an important medicine for dropsy, an ailment characterized by swelling of the limbs and torso, which we now know is due to inadequate pumping action of the heart.

There is certainly little in Withering's upbringing to suggest that he would one day produce one of the first modern studies in ethnobotany by interviewing a folk healer and carefully investigating the pharmacological activity of the plants she used. Like many modern premedical students, Withering was not enamored of the necessity of learning botany. In a letter to his parents Withering described his botany professor at Edinburgh, John Hope, as quite dull:

> The Botanical Professor gives annually a gold medal to such of his pupils as are most industrious in that branch of science. An incitement of this kind is often productive of the greatest emulation in young minds, though, I confess, it will hardly have charm enough to banish the disagreeable ideas I have formed of the study of botany.

Withering's botanical interest lay dormant until 1775, when he was smitten with Helen Cookes, an aspiring artist who liked to paint flowers. Young Withering, eager to please, collected plants for her to sketch. During this romantic interlude, the plants captured Withering's imagination. Although he continued his practice of medicine, he later published several texts on botany and was elected a Fellow of the Linnean Society of London. Thus trained in both medicine and botany, Withering was well prepared to make the most important ethnobotanical discovery of his age:

> In the year 1775, my opinion was asked concerning a family receipt for the cure of the dropsy. I was told that it had long been kept a secret by an old woman in Shropshire, who had sometimes made cures after the more regular practitioners had failed. . . . This medicine was composed of twenty or more different herbs, but it was not very difficult for one conversant in these subjects, to perceive, that the active herb could be no other than Foxglove.

The retention of fluid that swelled the dropsy patient's body was clearly alleviated by the administration of foxglove, but the connection between dropsy

English folk healers prescribed the foxglove plant, *Digitalis purpurea,* to treat dropsy, a condition caused by inadequate pumping action of the heart.

A portrait of William Withering, the English physician and botanist who, while investigating the folk use of plant mixtures to treat dropsy, discovered that these mixtures shared the common element foxglove, the plant that he holds in his hand.

and inadequate pumping action of the heart was not properly understood in Withering's day. In a brilliant insight, Withering observed that foxglove "has a power over the motion of the heart, to a degree yet unobserved in any other medicine." Withering foresaw that "this power may be converted to salutary ends." He began prescribing foxglove for cases of dropsy, "but I gave it in doses very much too large."

Part of his problem was standardizing the dosage from ground leaves:

> These I had found to vary much as to dose, at different seasons of the
> year; but I expected, if gathered always in one condition, viz. when it
> was flowering late, and carefully dried, that the dose might be ascer-
> tained as exactly as that of any other medicine; nor have I been dis-
> appointed in this expectation.

Withering soon began to prescribe infusions of the leaves (which he made by
steeping the leaves in water) and, later, ground leaf powder. By any standard,
foxglove as administered by Withering was an astonishingly successful treatment for dropsy. When J. K. Aronson at Oxford recently reanalyzed data from
the cases Withering carefully recorded, he found a success rate of between 65
and 80 percent.

Powdered foxglove leaf is still prescribed in tablet or capsule form to treat
congestive heart failure. The Latin name of the foxglove genus, *Digitalis*,
has been affixed to this crude drug as well as to the cardiac glycosides isolated
from foxglove in the early twentieth century. Cardiac glycosides are steroidal
compounds (naturally occurring compounds with a characteristic 17-carbon
skeleton) with attached sugars, and they are so named because of their powerful action on the heart. These drugs are useful because they increase the force
of heart contractions and allow the heart more time to rest between contractions. More than 30 cardiac glycosides have been isolated from dried foxglove
leaves, including digitoxin and digoxin. Neither of these drugs has ever been
commercially synthesized; both are still extracted from dried foxglove leaves.
Each year over 1500 kilograms of pure digoxin and 200 kilograms of digitoxin

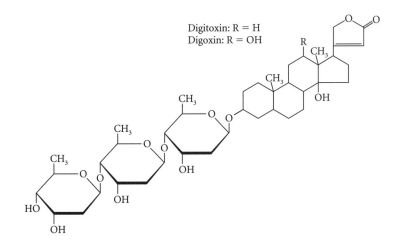

Digitoxin: R = H
Digoxin: R = OH

Digoxin and digitoxin, two important
heart medicines still extracted from the
foxglove plant, are classified as cardiac glycosides. "Cardiac" refers to their action on
the heart, while "glycosides" refers to the
linked sugar molecules shown at the lower
left of the structure.

are prescribed to hundreds of thousands of heart patients throughout the world.

Although the development of new pharmaceuticals has added to the repertoire of heart medicines, digitalis still saves many lives each year. Digitalis remains the drug of choice in the treatment of fast atrial fibrillation, a life-threatening condition characterized by unsynchronized contractions of the heart which reduce its pumping action. In perhaps the earliest demonstration of the power of the ethnobotanical approach to drug discovery, William Withering's willingness to consult "an old woman in Shropshire" eventually resulted in a drug of tremendous importance.

Withering's ethnobotanical study of a single plant, *Digitalis,* resulted in an important advance in medicine by reporting his investigation of the knowledge of folk healers in his own culture. His contemporary, the Swedish naturalist Carl Linnaeus, made no new drug discoveries but significantly advanced ethnobotany by studying the uses of plants in a culture other than his own. He also laid the foundation for systematizing all of botany. Before Linnaeus there was no unified scheme for plant nomenclature. Linnaeus codified the use of Latin binomials for plants, together with rules to distinguish between competing names invented by different botanists.

Linnaeus and Ethnobotany in Lapland

"I set out alone from the City of Uppsala on Friday, 12 May 1732, at 11 o'clock, being at that time within half a day of twenty-five years of age," Linnaeus wrote at the beginning of his journal in 1733. He was bound for Lapland, a region north of the Arctic Circle that is inhabited by the Lappish, or Sami, people. The Sami depend on semidomesticated reindeer herds for meat and hides. Linnaeus carefully recorded the plants the Sami use in their struggle to deal with the harsh Arctic environment. His journal, filled with sketches of the Sami people and notes on their uses of plants, exemplifies the detailed observation that continues to characterize ethnobotanical studies:

> June 5. The bountiful provision of nature is evinced in providing mankind with bed and bedding even in this savage wilderness. The *Polytrichum* prolif. *maximum* [moss] grows copiously in the damp forests and is used for this purpose. They choose the starry headed plants, out of the tufts of which they cut a surface as large as they please for bed or bolster.

June 13. In the neighborhood grows *Pinguicula*. When the inhabitants of these parts once procure this plant, they avail themselves of it throughout the whole year, and use it as a kind of rennet until the return of spring.

After his return to Uppsala in September, Linnaeus became the talk of the town by displaying and wearing Lappish garments and giving empathetic accounts of a people who were largely unknown to his countrymen. His published journal of his travels in Lapland, showing how the Lappish people used plants for a variety of needs, achieved wide circulation. His observation of the use of the leaves of the insectivorous plant *Pinguicula* [Lentibulariaceae] to curdle milk, for example, demonstrated how people had exploited for a new purpose an enzyme used by the plant for the digestion of insects. His work is of such exceptional detail that it is being reexamined today at the University of Uppsala, where investigators are extending his studies of Sami ethnobotany.

Richard Evans Schultes and Ethnobotany in the Amazon

Linnaeus pioneered several techniques that went far beyond William Withering's. On his travels in search of ethnobotanical knowledge Linnaeus learned exotic languages. He traveled alone or with only a few companions and with a minimum of gear. In the field Linnaeus ate indigenous foods and learned to use plants as the indigenous peoples used them. Most important, Linnaeus established a deep rapport with the people he studied. Two centuries later, all of these characteristics would be seen again in the ethnobotanical work of Richard Evans Schultes at Harvard University.

As an undergraduate, Schultes chose to write about the peyote cactus (*Lophophora williamsii* [Cactaceae]) for his senior thesis. His adviser, Oakes Ames, insisted that Schultes have first-hand field experience with the plant, so in 1937 Schultes journeyed to Oklahoma to study with the Kiowa Indians and learn about their ceremonial use of this tiny cactus. Young as he was, he returned with one of the most careful analyses of folk use of a hallucinogenic plant that had ever been written. Ames suggested that Schultes continue his ethnobotanical interests while he pursued a doctoral program.

Schultes' doctoral research took him to Mexico in search of the sacred mushroom of the Aztecs, *Teonanacatl* (*Panaeolus campanulatus* and related species). Although the mushroom's existence had been rumored, Schultes was the first botanist to record the rituals and beliefs surrounding this sacred mushroom. After receiving his Ph.D. in 1941, Schultes journeyed to investigate the ethno-

After his journey to Lapland, pioneer ethnobotanist Carl Linnaeus sometimes wore Lappish garments to demonstrate the ways of the Sami people for his students in Uppsala University.

botany of the tribal peoples of the northwest Amazon. Returning from the Colombian Amazon to Bogotá, Schultes heard some alarming news: the United States had entered World War II. Wishing to return home immediately, Schultes went to the U.S. embassy, only to be told that the government had different plans for him: they wanted him to return to the rain forest.

Few American botanists had extensive experience in the Amazon area, and Schultes' expertise was sought by the Allies when the United States entered World War II. With the fall of Burma, Malaysia, and Indonesia to the Japanese, the Allies' access to rubber from plantations in those areas vanished. With no time to engage in an extensive planting program, it became important to determine if the rubber the Allies needed could be supplied from wild trees in the rain forests of the Amazon.

Schultes was assigned to determine the density of rubber trees (*Hevea brasiliensis* [Euphorbiaceae]) in the rain forest and to see if the local Indians could harvest the rubber. Although on an important mission for the military in a little-known part of the world, Schultes chose to conduct the survey as he had done all of his previous research: he traveled alone in a canoe with no weapons and a minimum of gear, relying solely on his ability to interact with the local

This photograph taken of Richard Evans Schultes in the field shows him in the northwest Amazon basin of Colombia preparing herbarium specimens of locally used plants.

people to secure safe passage, shelter, and food. While he conducted this survey, he simultaneously carried out the ethnobotanical studies that would become the foundation for his book *The Healing Forest,* coauthored with Robert Raffauf and published in 1991. Schultes dutifully reported his findings to the military, and when the war ended, he chose to continue to work in the Amazon.

Over a 14-year period of continued residence in the northwest Amazon, Schultes assembled a collection of more than 25,000 plant specimens, many of which had economic value. He worked with many Amazonian tribes, identifying dozens of their hallucinogens and hundreds of their medicinal and toxic plants. Schultes immersed himself in village life, spending months at a time in the areas he was studying, returning to the Colombian river village of Mitú only to send out mail and reports and replenish his supplies. As a pioneer of the participant-observer method, Schultes went beyond the typical observer stance of most anthropologists of the day and participated in indigenous rituals involving plants. Perhaps because of this approach, Schultes and his students have become close to the indigenous peoples they work with. The writings of Richard Evans Schultes are known for the profound respect they express for the people, their culture, and their stewardship of the environment.

Schultes was an early proponent of an interdisciplinary approach in ethnobotany. Intrigued by the psychoactive principles of the Indians' hallucinogens, he began to collaborate closely with chemists and pharmacologists such as Albert Hofmann, the discoverer of LSD. Together they launched an investigation of the psychoactive principles of the Mexican marigold *Tagetes lucida* [Asteraceae]. Rumors of a hallucinogenic marigold had long been discounted by the pharmacological community, but Schultes' careful ethnobotanical work combined with Hofmann's painstaking chemical analysis of plant samples showed that the active component was astonishingly similar in structure to LSD-25.

Through his diligence and his large circle of collaborators, Schultes was able to transform the Botanical Museum at Harvard University into a major international center of ethnobotany. There as graduate students in the late 1970s we were both deeply influenced by his enthusiasm and his deep humanity.

Albert Hofmann, former director of research for the Department of Natural Products of Sandoz Pharmaceuticals in Basel, Switzerland. Hofmann discovered the powerful effects of LSD in April of 1953, and later collaborated with Richard Evans Schultes in an interdisciplinary study of the Mexican marigold *Tagetes lucida.*

Ethnobotany Today

With Professor Schultes as a model, we have spent extended periods in remote villages in the tropics studying the uses of plants by indigenous peoples. He taught that ethnobotanists slowly and meticulously learn about the plants the inhabitants use, catalogue their knowledge about the useful ones and the poisonous ones, and collect plants for study and possible cultivation. They accept lengthy stays in villages, hundreds of hours of patient observation and experimentation, and, above all, the repetitious but critical work of pressing and drying plant samples, even in monsoon rains and oppressive heat. To colleagues more familiar with test tubes and nuclear magnetic resonance spectroscopy than with notebooks and plant presses, this approach may seem far from scientific. Ethnobotany, like anthropology, geology, or even astronomy, is at its core an observational rather than an experimental science, but it is just as rigorous as, say, chemistry. In fact, modern ethnobotanists use sophisticated chemical analyses to elucidate the bioactive components of the plants they collect; incorporate statistical polling techniques to acquire and analyze data on plant uses; document with audiotape native ethnotaxonomies (indigenous classification systems); analyze genetic variation in native crops; and use carbon dating techniques to illuminate prehistorical uses of plants. Because of the need to incorporate techniques and insights from other fields, relatively few established investigators in ethnobotany were trained primarily to be ethnobotanists. Most have approached the field indirectly, from anthropology, linguistics, natural-product chemistry, pharmacognosy (the study of natural drugs and their constituents), systematic botany, or plant ecology. While this melding of many different approaches and disciplines has enriched the field, the lack of a clear consensus on such basic issues as disciplinary goals and study methodologies has hampered the development of a unified approach.

This poster photographed on a street in Cuzco, Peru, illustrates the commercial pressures that are causing many indigenous farmers to switch to modern varieties of agricultural crops. The displacement of traditional varieties is leading to the loss of genetic diversity. Vast regions are being left vulnerable to famine, since traditional varieties are often better adapted to survive the stresses of local conditions. Ethnobotanists help local people to develop strategies for preserving this valuable material.

The lack of a unified approach has led to a paucity of comparative ethnobotanical studies—studies that not only examine different uses of the same types of plants in different cultures, but also compare the ways plants figure in different worldviews. Few papers published by the leading ethnobotanical journals, *Economic Botany,* the *Journal of Ethnopharmacology,* and the *Journal of Ethnobiology,* comprehensively grapple with literature beyond the immediate topic at hand. Other events are compounding the problem. First, the loss of indigenous knowledge systems is accelerating throughout the world. Most ethnobotanists are alarmed by the rate at which plant lore is disappearing. As traditional peoples become increasingly Westernized, much of the richness of their traditions disappears. Plant lore that is passed down to only selected members of a community appears to be particularly susceptible to such cultural erosion. And second, the worldwide loss of plant biodiversity is accelerating as well. Knowledge of the uses of plants from threatened habitats appears to be especially vulnerable to loss.

In this book we will use selected cases to demonstrate (1) that plants have played a major role in determining the trajectories of modern culture, (2) that the wisdom of indigenous peoples can not only provide insight into the human condition but also enrich Western cultures, and (3) that conservation of plant biodiversity and indigenous plant lore is in the interest of the world community. On these central issues we believe most ethnobotanists are in agreement.

There has never been a more exciting time in the history of ethnobotany. Recent advances in molecular biology and analytical chemistry have equipped ethnobotanists with techniques and tools that our predecessors could scarcely imagine. The breadth of questions being pursued is also new: How did indigenous peoples choose plants to build vessels that would carry them across thousands of miles of ocean? Why did the ancient inhabitants of the Colorado Plateau abandon their homeland? How did different cultures find plants that produce almost identical psychoactive compounds? Ease of air travel, international collaboration, and interdisciplinary approaches have facilitated the pursuit of answers to these questions both in the laboratory and in the field. Recent methodological and conceptual advances have borne the most fruit in the search for novel pharmaceutical compounds. Understandably, the search for new drugs among plants used as traditional medicines has become the type of ethnobotanical investigation that has received the most public attention. Few people realize, however, that this is one of the oldest forms of ethnobotanical research known.

This interior of an Ayurvedic medicine shop—in the town of Kottakkal, Kerala State, in southern India—shows the vast array of products available for this system of traditional medicine.

Plants
That Heal

Walk into any pharmacy in the United States, Canada, or Western Europe and ask to examine any bottle of prescription medicine chosen at random. There is a one in four chance that the medicine you hold in your hand has an active ingredient derived from a plant. Most of these plant-derived drugs were originally discovered through the study of traditional cures and folk knowledge of indigenous peoples—the ethnobotanical approach. The pharmacognosist Norman Farnsworth, of the University of Illinois, estimates that 89 plant-derived drugs currently prescribed in the industrialized world were discovered by studying folk knowledge.

Consider again William Withering's discovery of digitalis in the eighteenth century. Withering's discovery depended on the same sequence of steps that lead to success in modern ethnobotanical drug discovery programs: (1) folk knowledge of a plant's possible therapeutic activity accumulates; (2) a healer uses the plant for her patients; (3) the healer communicates her knowledge to a scientist; (4) the scientist collects and identifies the plant; (5) the scientist tests extracts of the plant with a bioassay (a preliminary screen for the desired pharmacological activity); (6) the scientist isolates a pure compound by using the bioassay to trace the source of the activity in the plant extract; and (7) the scientist determines the structure of the pure substance.

In the case of digitalis, knowledge concerning the use of the foxglove accumulated in the traditional healing systems of the British people as recorded in Gerard's herbal of 1597. In the eighteenth century a healer (the "old woman" in Shropshire) used the plant to treat dropsy, an ailment caused by inadequate pumping action of the heart. Hearing of the therapy's success, Withering interviewed the healer, who shared her remedy with him. Withering identified and collected foxglove and then tested extracts of its leaves by the only means of bioassay available to him: he gave it to his patients. Today it is considered unethical to screen untested substances on human populations, but in the eighteenth century Withering had no access to laboratory animals with dropsy. By the end of the nineteenth century, chemists were able to separate the components of the foxglove leaves through extraction with solvents (a process known as fractionation) and show that the observed cardiotonic effect was largely due to the cardiac glycosides. They determined the structures of the glycosides and named two of the most active ones digoxin and digitoxin.

Although William Withering is usually credited with the discovery of digitalis, there was a long interval between his record of the folk use of foxglove leaves and the production of pure digoxin and digitoxin on an industrial scale. This gap, extending nearly a century and a half, was bridged by the work of many other chemists, pharmacologists, and medical workers. Modern ethnobotanical research programs are designed to compress a process that once took centuries. Yet many factors far removed from scientific considerations influence the rate at which drugs are discovered and produced. Few, if any, drugs, are produced merely as research curiosities. Strong economic, social, and even political incentives are needed to fuel the arduous processes of discovering and developing drugs. The spread of new diseases, often as a result of human colonization of new habitats, and restricted access to drug supplies in time of war have often spurred innovation and resourcefulness. The power of social, political, and economic conditions to provide strong incentives for drug production can be clearly seen in the history of a drug very different from digitalis: quinine.

"Peruvian Bark" and the Discovery of Quinine

Quinine is an odorless white powder with an extraordinarily bitter taste. It is useful against malaria, the disease caused by mosquito-borne transmission of the protozoan *Plasmodium*. It is also used to treat cardiac arrhythmias. Although we often think of malaria as a tropical disease, it once was a serious health hazard in such temperate cities as Washington, St. Louis, London, and Rome. (Miasma, a nighttime vapor that rises from swamps and other wet places, was believed to cause malaria before mosquitoes were found to be the vector.)

Quinine is valued as a flavoring agent as well as a medicine. Most quinine imported into the United States is used to flavor tonic water. Yet the taste for the flavor of tonic water can be traced to the use of quinine as an antimalarial compound by British colonial forces in India. Since the blood serum level of quinine needed to protect against malaria is rather low, a daily gin and tonic may indeed have some small medicinal value in malarial regions.

The story of the West's discovery of quinine begins in the late sixteenth and early seventeenth centuries, during the conquest of the Inca empire in Peru. The Spanish invaders became aware of a montane rain forest tree used by the Indians to treat fevers. A Spanish legend says that a soldier who was suffering from a bout of malaria in the wilderness drank the dark-brown water in a pool into which quinine trees had fallen. He then went to sleep, and when he awoke, he found that his fever had disappeared. Concluding that the bitter brown water was a "tea" made from the steeped stems and bark of the tree, he spread the word about its power to reduce his fever. Another Spanish legend tells of Indians who observed that sick animals came to drink at the tepid pools around which great stands of quinine trees grew.

In 1633 a Jesuit priest named Father Calancha described the healing properties of the tree in *The Chronicle of St. Augustine:*

> A tree grows which they call the fever tree in the country of Loxa, whose bark is the color of cinnamon. When made into a powder amounting to the weight of two silver coins and given as a beverage, it cures the fevers and . . . has produced miraculous cures in Lima.

Jesuits throughout Peru began using the bark to prevent and treat malaria. In 1645 Father Bartolomé Tafur took some bark to Rome, where its use spread among clerics. Cardinal John de Lugo wrote a leaflet to be distributed with the bark. Because the miraculous "Peruvian bark" was so widely used, not a single

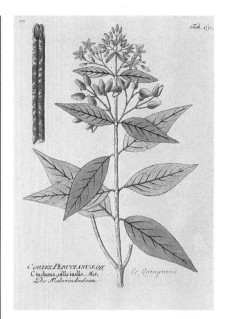

A flowering branch of one of the species of *Cinchona* used commercially to produce quinine. A strip of bark, harvested from the stem of the tree, is shown at the upper left. It is this bark that is the source of quinine used in pharmacy.

participant in the papal conclave of 1655 died of malaria—the first time in recorded history that a Roman convocation had been so spared. Peruvian bark was introduced to England as early as 1654, but British Protestants were reluctant to try a Catholic concoction. Oliver Cromwell, who refused to be "Jesuited" with the bark, died of malaria in 1658.

In 1670 a young apothecary named Robert Talbor gained fame in London by curing malaria with a secret formula. Talbor belittled the Peruvian bark, warning the public to "Beware of all palliative Cures and especially that known by the name of 'Jesuit's powder'." After Talbor's secret formula cured Charles II's malaria, the king sent him to the French court, where he successfully treated the ailing son of Louis XIV. The French king paid 3000 gold crowns for Talbor's secret, which Talbor stipulated could be published only after his death. It was then discovered that the "secret" was Peruvian bark.

Despite the fame of Peruvian bark, its botany remained unknown: no botanist had ever published a description or drawing of the tree from which it came because it grew in high rain forests in the Andes. In 1735 a French botanist named Joseph de Jussieu traveled to South America and after many travails found and described the tree, a small member of the Rubiaceae, or coffee family, that grows in the understory. In 1739 the Swedish taxonomist Carl Linnaeus named the genus *Cinchona*, a misspelling of the name of a Spanish countess who, legend claimed, had been healed by the bark.

In 1820 the French chemists Joseph Pelletier and Joseph Caventou isolated the alkaloid quinine from the bark and were awarded 10,000 francs by the Paris Institute of Science. Yet, although the purified alkaloid quinine had been discovered, no one could synthesize it. Quinine producers therefore continued to rely on massive supplies of bark collected from wild *Cinchona* trees. In 1880 Colombia alone exported 6 million pounds to Europe, all collected from undomesticated forest trees. The export value of *Cinchona* bark was so great that Bolivia, Colombia, Ecuador, and Peru attempted to maintain a tight monopoly on production by prohibiting the export of seeds or living plants. But the temptation to break the South American monopoly proved irresistible, and in 1852 Justus Hasskarl, the director of a Dutch botanical garden in Java, began a secret mission to smuggle *Cinchona* seeds out of South America.

Hasskarl's plan was exposed by a German newspaper, but the following year he entered South America under an assumed name and traded a bag of gold to an official for *Cinchona* seeds. Returning to Java with his botanical bounty, he was immediately knighted by the Dutch government. As the trees matured, however, jubilation changed to dismay, for the quinine content of their bark proved to be disappointingly low. Clearly *Cinchona* trees differed

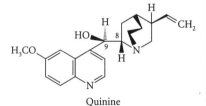

Quinine

from strain to strain in the amount of alkaloids they produced. A second attempt to collect *Cinchona* seeds would be needed to produce a viable industry in Java.

Another opportunity to establish a Dutch quinine industry was inadvertently provided by an Australian, Charles Ledger, in 1861. Ledger had tried on several occasions to collect *Cinchona* seeds but was bewildered by the diversity of this genus—there are 40 species, and each species has countless strains. As it happened, seeds that Ledger had sold the British government produced trees with very low quinine content. However, Ledger eventually prevailed upon an Aymará Indian, Manuel Incra, to smuggle seeds from a species of *Cinchona* tree in Bolivia that was reputed to have high quinine content. Upon discovery of this infringement, the Bolivian government tortured Incra to death. Ledger traveled to Europe and attempted to sell his seeds to the British government. Because of the low alkaloid content of the *Cinchona* plants Ledger had provided earlier, the British government refused to buy any, but a pound of Ledger's seeds eventually found their way to the Dutch government. The Dutch paid the equivalent of $20 for them and sent them to Java to be planted. It was arguably the best $20 investment made in history.

As the trees matured, the Dutch were astonished to discover that their bark had a record alkaloid content of 13 percent. As the new alkaloid-rich strain came into production in Java, the harvesting of wild plants, which typically had lower quinine content, in South America waned. By 1930 the Dutch plantations in Java produced 22 million pounds of bark, yielding 97 percent of the world's quinine.

Yet eventually, this Dutch near-monopoly on quinine inadvertently threatened the stability of Western democracy. In 1940 the German army seized the entire European repository of quinine when it captured Amsterdam. When the Japanese conquered Indonesia in 1942, the United States and its allies were virtually without quinine supplies. There was a small *Cinchona* plantation in the Philippines, but it, too, fell to the Japanese only weeks after they annexed Java. The last Allied plane to leave the Philippines before the islands capitulated to the Japanese contained a singularly precious cargo: together with key Philippine personnel, the aircraft carried 4 million tiny *Cinchona* seeds. These were flown directly to Maryland. When they had germinated, they were sent to Costa Rica for planting. Although the evacuation of this *Cinchona* germplasm (genetic material) from the Philippines was carried out with considerable foresight and valor, there was little hope that the resultant trees could mature quickly enough to meet the urgent wartime need for quinine. More than 600,000 U.S. troops in Africa and the South Pacific had contracted malaria, and the average mortality

Smithsonian Institution botanist Raymond Fosberg, shown here in the field in 1948, was recruited by the U.S. government to resecure supplies of *Cinchona* bark after Japan seized all known cultivated sources at the onset of World War II.

rate was 10 percent. Since more U.S. soldiers were dying from malaria than from Japanese bullets, the lack of *Cinchona* bark immediately became a serious national security issue.

A few weeks after the fall of the Philippines, the botanist Raymond Fosberg received an unusual delegation in his office at the Smithsonian Institution. As one of the few North American tropical biologists, he was asked by the U.S. Board of Economic Warfare to carry out a mission of highest priority. Together with several other U.S. botanists, he was to travel immediately to South America, recollect all known *Cinchona* species, secure a massive supply of *Cinchona* bark for shipment to the United States, and establish local plantations of *Cinchona* trees. Fosberg was to obtain, if possible, millions of pounds of bark for immediate shipment to the Merck pharmaceutical plant in New Jersey.

Fosberg was placed in charge of the survey in Colombia. So that chemical assays of any bark collected could be carried out quickly, the U.S. government set up field laboratories in Bogotá, Colombia; Quito, Ecuador; Lima, Peru; and La Paz, Bolivia. Since the exact locality of the original sixteenth-century collections was unknown, Fosberg traveled for months with local assistants through remote forests interviewing Indians and searching for *Cinchona* species. When he located a large stand of the trees, he had to arrange for local people to harvest the bark, dry it under difficult tropical conditions, and then transport it down a mule or foot trail to the nearest road or river. If no trail was available, Fosberg worked with the Indians to cut an airstrip in the jungle so that the bark could be flown out.

In the course of these expeditions, Fosberg and his colleagues learned a great deal about *Cinchona* biology. In time they were able to project how much dried bark a tree of a particular size would yield. A tree whose stem measured 2 inches in diameter was found to yield 1 pound of bark; a tree whose stem measured 26 inches in diameter yielded 255 pounds.

The emergency explorations had mixed success. In 1943 and 1944, Fosberg and his colleagues secured 12.5 million pounds of *Cinchona* bark for the Allies' war effort. Yet they never did locate the quinine-producing species, *Cinchona ledgeriana*, that had made the Java plantations so productive. Meanwhile, Allied chemists searched for quinine substitutes, but synthetic antimalarial drugs lacked the efficacy of real quinine and produced such unpleasant effects as nausea, diarrhea, and yellowing of the skin, which made them unpopular with American soldiers.

As the war continued, Fosberg persisted in his search for *Cinchona* species but soon had to confront a problem far more serious than the elusiveness of a tree: he became aware that he himself was being hunted. As Fosberg tells the

During the collection effort mounted at the time of World War II, once bark was harvested from wild stands of *Cinchona,* it was quickly dried in the sun to preserve its quinine content. This photograph shows *Cinchona* bark being dried after collection in the forests of Ecuador in the summer of 1944.

story, he had just checked into a run-down hotel in a remote Colombian outpost when he heard German voices coming from the room below. Late that night a knock came at his door. Fosberg opened it to find himself staring into the faces of two Nazi agents, who explained they knew who he was and what he was doing there. They had been on his trail for several weeks. Would the U.S. government, they asked, be interested in buying a large quantity of pure quinine that they had smuggled out of Germany? Relieved, Fosberg struck a deal and returned to the United States with German quinine, which quickly and quietly found its way to the Pacific theater.

After the war, synthetic antimalarial drugs such as Maloprim and Fansidar reduced the need for quinine. But the utility of quinine to treat certain heart arrhythmias and its commercial value as a bitter flavoring agent suggest that this bark, which first made its way from Peru to the royal courts of Europe, will remain an important botanical commodity for years to come.

A Successful Approach to Drug Discovery

As the discoveries of digitalis and quinine bear witness, the ethnobotanical approach to drug discovery has been spectacularly successful. The table on pages 34 – 35 lists 50 drugs prescribed in North America and Europe that were derived from ethnobotanical leads. Most of them were discovered from leads known to Western science for decades. We think of William Withering as the pioneer of cardioactive drugs with his discovery of digitalis, for instance, but in 1597 Gerard noted that the sea squill (*Drimia maritima* [Liliaceae]) "is given to those that have the dropsie." Since then, the cardiotonic drug Proscillaridin has been derived from *D. maritima.*

Aspirin is yet another drug developed from a plant. The European herb called queen of the meadow, *Filipendula ulmaria* [Rosaceae], which is referred to in some older literature as *Spiraea ulmaria,* has long been used in folk medicine to treat pain and fevers and as an antiseptic. In 1597 Gerard wrote that the roots of this plant, "when boiled in wine and drunken, are useful against all pains of the bladder." In 1839 salicylic acid was isolated from the flower buds of *F. ulmaria.* The pure compound rapidly came into widespread use as a pain reliever, but it frequently caused gastric upset. Then in 1899 the Bayer company began to market a synthetic derivative, acetylsalicylic acid, which had higher pharmacological activity and fewer side effects. They called their new compound "aspirin"—"a" for "acetyl" and "spirin" for *Spiraea,* the plant from which salicylic acid had originally been derived. Salicylic acid also occurs in members of the willow family, the Salicaceae. The ancient Greeks and North American Indians used the bark of the willow genus *Salix* to relieve pain.

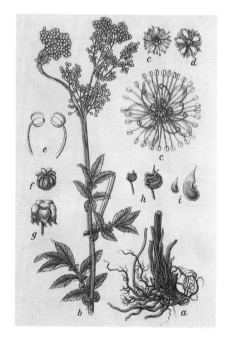

Aspirin
(Acetylsalicylic acid)

Queen of the meadow, *Filipendula ulmaria,* was long used in folk medicine to treat pain and fevers and as an antiseptic. It is the original source of salicylic acid, the precursor of modern aspirin.

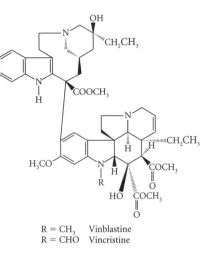

R = CH₃ Vinblastine
R = CHO Vincristine

A healer's apprentice in Belize, Bertha Waight, examines rosy periwinkle, *Catharanthus roseus,* as part of her studies. The rosy periwinkle originated in the forests of Madagascar, but is now found growing wild in many areas of the tropics. Two potent chemicals derived from the leaves of this plant, vincristine and vinblastine, are the drugs of choice to treat certain cancers.

The fifty ethnobotanically derived drugs listed in the table on pages 34–35 pose an interesting question: Can folk wisdom still point the way to new drugs? A decade or so ago, the story of Withering's discovery of digitalis might have been regarded as a historical anecdote of little relevance to contemporary drug discovery programs, even though such discoveries have continued well into the twentieth century. Perhaps the most significant discovery was that of the vinca alkaloids, vincristine and vinblastine, in the rosy periwinkle, *Catharanthus roseus* [Apocynaceae]. These alkaloids are used around the world for the treatment of pediatric leukemia and Hodgkin's disease. The rosy periwinkle was discovered in a collection of 400 medicinal plants that scientists at Eli Lilly screened against cultures of P-38 mouse-cell leukemia. In the laboratory, rosy periwinkle killed leukemia cells. The active components, vincristine and vinblastine, occur in such low concentrations that more than 250 kilograms of leaves are needed to make a single 500-milligram dose. Although the plant is unlikely to be useful against leukemia in a folk setting, it was indeed a healer's claim that the plant was effective against diabetes that led scientists to investigate it.

More recently the effort to make use of folk knowledge in the search for novel pharmaceuticals has increased throughout the world. The task is daunting. Modern searches for bioactive molecules typically make use of expensive

Fifty Drugs Discovered from Ethnobotanical Leads

Drug	Medical Use	Plant Species	Family
Ajmaline	Heart arrhythmia	*Rauvolfia* spp.	Apocynaceae
Aspirin	Analgesic, inflammation	*Filipendula ulmaria*	Rosaceae
Atropine	Ophthalmology	*Atropa belladonna*	Solanaceae
Benzoin	Oral disinfectant	*Styrax tonkinensis*	Styracaceae
Caffeine	Stimulant	*Camellia sinensis*	Theaceae
Camphor	Rheumatic pain	*Cinnamomum camphora*	Lauraceae
Cascara	Purgative	*Rhamnus purshiana*	Rhamnaceae
Cocaine	Ophthalmologic anaesthetic	*Erythroxylum coca*	Erythroxylaceae
Codeine	Analgesic, antitussive	*Papaver somniferum*	Papaveraceae
Colchicine	Gout	*Colchicum autumnale*	Liliaceae
Demecolcine	Leukemia, lymphomata	*Colchicum autumnale*	Liliaceae
Deserpidine	Hypertension	*Rauvolfia canescens*	Apocynaceae
Dicoumarol	Thrombosis	*Melilotus officinalis*	Fabaceae
Digitoxin	Atrial fibrillation	*Digitalis purpurea*	Scrophulariaceae
Digoxin	Atrial fibrillation	*Digitalis purpurea*	Scrophulariaceae
Emetine	Amoebic dysentery	*Cephaelis ipecachuanha*	Rubiaceae
Ephedrine	Bronchodilator	*Ephedra sinica*	Ephedraceae
Eugenol	Toothache	*Syzygium aromaticum*	Myrtaceae
Gallotanins	Hemorrhoid suppository	*Hamamelis virginiana*	Hamamelidaceae
Hyoscyamine	Anticholinergic	*Hyoscyamus niger*	Solanaceae
Ipecac	Emetic	*Cephaelis ipecacuanha*	Rubiaceae
Ipratropium	Bronchodilator	*Hyoscyamus niger*	Solanaceae
Morphine	Analgesic	*Papaver somniferum*	Papaveraceae
Noscapine	Antitussive	*Papaver somniferum*	Papaveraceae
Papain	Attenuates mucus	*Carica papaya*	Caricaceae

Fifty Drugs Discovered from Ethnobotanical Leads *(Continued)*

Drug	Medical Use	Plant Species	Family
Papaverine	Antispasmodic	*Papaver somniferum*	Papaveraceae
Physostigmine	Glaucoma	*Physostigma venenosum*	Fabaceae
Picrotoxin	Barbiturate antidote	*Anamirta cocculus*	Menispermaceae
Pilocarpine	Glaucoma	*Pilocarpus jaborandi*	Rutaceae
Podophyllotoxin	Condylomata acuminata	*Podophyllum peltatum*	Berberidaceae
Proscillaridin	Cardiac malfunction	*Drimia maritima*	Liliaceae
Protoveratrine	Hypertension	*Veratrum album*	Liliaceae
Pseudoephedrine	Rhinitis	*Ephedra sinica*	Ephedraceae
Psoralen	Vitiligo	*Psoralea corylifolia*	Fabaceae
Quinidine	Cardiac arrhythmia	*Cinchona pubescens*	Rubiaceae
Quinine	Malaria prophylaxis	*Cinchona pubescens*	Rubiaceae
Rescinnamine	Hypertension	*Rauvolfia serpentina*	Apocynaceae
Reserpine	Hypertension	*Rauvolfia serpentina*	Apocynaceae
Sennoside A,B	Laxative	*Cassia angustifolia*	Caesalpiniaceae
Scopolamine	Motion sickness	*Datura stramonium*	Solanaceae
Stigmasterol	Steroidal precursor	*Physostigma venenosum*	Fabaceae
Strophanthin	Congestive heart failure	*Strophanthus gratus*	Apocynaceae
Teniposide	Bladder neoplasms	*Podophyllum peltatum*	Berberidaceae
THC	Antiemetic	*Cannabis sativa*	Cannabaceae
Theophylline	Diuretic, asthma	*Camellia sinensis*	Theaceae
Toxiferine	Surgery, relaxant	*Strychnos guianensis*	Loganiaceae
Tubocurarine	Muscle relaxant	*Chondrodendron tomentosum*	Menispermaceae
Vinblastine	Hodgkin's disease	*Catharanthus roseus*	Apocynaceae
Vincristine	Pediatric leukemia	*Catharanthus roseus*	Apocynaceae
Xanthotoxin	Vitiligo	*Ammi majus*	Apiaceae

molecular biology assays in attempts to identify specific interactions along biochemical pathways of particular disease targets. Often an entire biochemical pathway depends on a key enzyme; if the enzyme can be inactivated by a drug, the pathway will be blocked and the disease averted. Once the underlying biochemistry is understood, automated screens of thousands of substances, including plant extracts, can be rapidly conducted. Should activity be found, scientists then work to isolate, purify, and determine the structure of the bioactive molecule. Since the funds available for such searches are limited, it is clear that not every one of the 250,000 different species of flowering plant species can be carefully examined. Indeed, since the beginning of modern pharmacology, less than $\frac{1}{2}$ of 1 percent of the species of flowering plants have been exhaustively studied to determine their chemical composition and medical potential.

A problem more difficult than the financial one has been a deep-rooted prejudice in the pharmacological community against ethnobotanical searches. Although ethnobotanical approaches to drug discovery are of historical significance, pharmaceutical firms in the 1960s and 1970s believed that new approaches, incorporating techniques from molecular biology and computer-assisted drug design, had superseded folk knowledge as a potential source of new pharmaceuticals. Ethnobotanical approaches such as that employed by William Withering came to be regarded as antiquated in comparison with computer-assisted design of pharmaceuticals.

A deeper reluctance to explore indigenous knowledge systems may be attributable to cultural prejudice dating to the years when the Western powers reigned over colonies. During the period of colonial expansionism, Western medicine "was taken as a prime exemplar of the constructive and beneficial effects of European rule," writes David Arnold, a historian of science at the University of Manchester. "Thus Western medicine was to the imperial mind . . . one of its most indisputable claims to legitimacy." Since Western medicine was regarded as prima facie evidence of the intellectual and cultural superiority of Europeans, the figure of the medicine man or shaman was often viewed as inimical to social and cultural progress. Indeed, the pejorative term "witch doctor" has come to stand for savagery, superstition, irrationality, and malevolence.

Why, then, should scientists spend considerable time and effort to seek out and study with the very healers that Western culture has so long held in contempt? As often occurs in science, the pendulum is beginning to swing back. Plant-based pharmaceuticals are again considered worth pursuing, for several reasons. First, we are becoming increasingly aware of the loss of biodiversity throughout the world, a loss that may preclude future discoveries of plant-based pharmaceuticals. Second, new molecular tools for screening novel substances have greatly accelerated the pace of research. Not long ago individual laboratory animals had to be injected with plant extracts; today an automated bioassay can screen hundreds of extracts in a few hours. As a result, both the amount of plant extract used and the time needed to determine any bioactivity have been reduced significantly. Third, appreciation of the sophistication of indigenous knowledge systems has grown. Claims that a traditional remedy really works are no longer dismissed out of hand—there are simply too many historical precedents of major drug discoveries rooted in folk medicine. One new pharmaceutical company, Shaman Pharmaceuticals, Inc., was established in 1989 to prospect for therapies based entirely on ethnobotanical leads.

Not all plant-derived pharmaceuticals, however, are products of ethnobotanical research. Traditionally, two different approaches, random and targeted, have been attempted in the selection of plant species for investigation. In random plant selection programs, a broad net is cast and plants are collected from a given region and screened without regard to their taxonomic affinities, ethnobotanical context, or other intrinsic qualities. Such searches have had consistently low success rates, although the National Cancer Institute (NCI) discovered taxol, an important drug used to treat breast and ovarian cancer, during a random screen.

Targeted selection programs can be of several types. In phylogenetic surveys, the close relatives of plants known to produce useful compounds are collected. In ecological surveys, plants that live in particular habitats or have certain char-

Medical doctor Thomas Carlson, on the left, and ethnobotanist Steven King, on the right, conducted field ethnobiomedical research in southeast Nigeria. Here, they are listening as Ester Madu, a traditional healer of Igbo culture, describes the use of a species for treating non-insulin-dependent diabetes mellitus, or type II diabetes. Ethnobotanist/physician teams are necessary to understand the range of information presented by traditional healers, who have extensive knowledge of both botany and medicine.

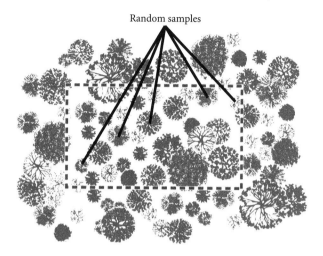

Random samples

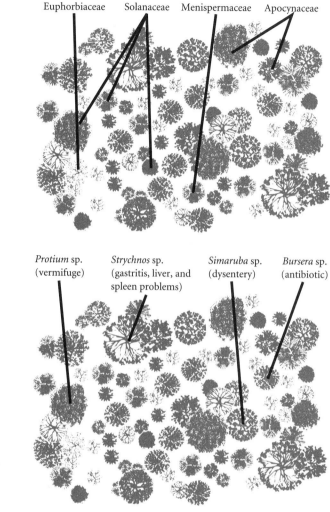

Euphorbiaceae   Solanaceae   Menispermaceae   Apocynaceae

Protium sp. (vermifuge)   *Strychnos* sp. (gastritis, liver, and spleen problems)   *Simaruba* sp. (dysentery)   *Bursera* sp. (antibiotic)

Strategies for collecting plants in the search for medicinal compounds include: (top left) taking samples of plants at random, (top right) taking samples from plants in families already known to contain plants with bioactive compounds, and (bottom) taking samples from plants used by traditional healers. The plants selected in the ethnodirected survey represented by the bottom panel were recommended by a Maya healer in Belize.

acteristics, such as immunity to predation by insects or molluscs, are selected. And in ethnobotanical surveys, plants used by indigenous peoples in traditional medicine are chosen for study.

The history of drug discovery and development seems to confirm that ethnobotanical screens of floras are far more likely to succeed than random screens. The ethno-directed sampling approach, as this methodology is called, has two primary components. The first is the cultural prescreen, in which indigenous peoples experiment with the plants in their environment, often over hundreds of generations, and identify those that are bioactive. The second component is a screen that the ethnobotanist consciously or subconsciously employs to determine which plants warrant further study. For example, claims of blowgun poisons or the mood-altering effects of plants in the dogbane family [Apocynaceae] are likely to arouse any ethnobotanist's interest, since many plants in this family are known to exhibit potent cardiotonic or psychoactive activity. Particular disease targets may also predispose the researcher to pay close attention to certain types of claims. In this age of AIDS and other viral diseases, any indication of possible antiviral activity for a plant used in traditional medicine is likely to be carefully evaluated.

This approach has been shown to increase the number of hits produced by *in vitro* studies (those carried out in test tubes) in comparison with the random approach. In broad-based *in vitro* screens, for example, Paul Cox, Rebecca Sperry, Lars Bohlin, and other colleagues at the University of Uppsala found 86 percent of the medicinal plant species in Samoa to show significant levels of pharmacological activity. Michael Balick, testing plant samples in a National Cancer Institute screen for anti-HIV activity, found that a small sample of "powerful plants" from an individual healer in a village in Belize, in Central America, initially gave four times as many hits in an HIV screen as a random collection. Steven King of Shaman Pharmaceuticals found that the type of pharmacological activity identified by *in vitro* bioassays corresponded with the activity identified by the indigenous healers 74 percent of the time. For many biological activities that have a corresponding folk use—antifungal, antibacterial, or hypoglycemic activity, for example—it appears that selecting plants that traditional healers use to treat these conditions will give higher levels of positive activity in the biological screens.

It is important to note, however, that not all indications of pharmacological activity lead to the discovery of new compounds. Many times this approach results in the isolation of compounds already known. It now appears that several plants that aroused Balick's interest in Belize, for instance, contain compounds that were already known to enhance the body's immune system. Because drug development programs like the one at the NCI are directed toward discovering new compounds, these plants, though active in the test tube studies, were not considered for further evaluation. The NCI's tests did make it clear, however, that healers were indeed able to identify plants with beneficial properties.

Still, the likelihood of success by the ethnobotanical approach may vary from culture to culture. Not all cultures are equally likely to use plants with significant pharmacological activity. Ethnobotanists tend to focus drug searches on cultures that have three characteristics: a cultural mechanism for the accurate transmission of ethnopharmacological knowledge from generation to generation, a floristically diverse environment, and continuity of residence in the area over many generations. Ethnobotanical data derived from cultures that display all three of these characteristics can be somewhat analogous to human bioassay data, particularly if people have been dosing themselves with the same plants for many generations. These people are likely to have identified any problems of lack of efficacy or acute toxicity over the years.

Ethnobotanists at Work in the Field

The drug discovery process, beginning with a plant used by a traditional healer and ending in a medicine used in a clinical setting, involves many disciplines and often takes many years to complete. Unlike the techniques used by pharmacologists or natural-product chemists, the skills required by the ethnobotanist are difficult to articulate in a book or classroom; for whereas a chemist's personality has little effect on the outcome of an experiment, the ethnobotanist's demeanor can have a direct impact on the success of the study.

Before a drug search can begin, the ethnobotanist must first obtain permission to conduct research from the national government. If the research is to be conducted in a foreign country, certain international protocols must be scrupulously observed. Under the Rio Treaty on Biodiversity, each signatory nation has sovereignty over all biodiversity within its boundaries. No plant sample that might result in discovery of a novel pharmaceutical compound can be removed without the country's written permission. Today Charles Ledger's smuggling of *Cinchona* seeds would be a crime punishable in the source country, Bolivia; in his home country, Australia; and in the recipient country, Indonesia.

Once the national government has given permission for the research, the ethnobotanist must obtain the permission of the village leaders. Because of issues of intellectual property rights, the ethnobotanist should negotiate in advance a fair and equitable return to the local people on any commercial development of a plant used in traditional medicine. The ethnobotanist must then meet and establish rapport with the village healers. The ability to secure and maintain the healers' trust is the single most important skill the ethnobotanist can have. Ethnobotanists establish rapport in several ways. Their first task is to learn the language of the people they study with. The use of an interpreter is seldom satisfactory, because healers employ special concepts and terms that most members of their culture do not know. "If you want to understand fully the ideas of sickness and health that underlie your healer's practices, his cate-

In Belize, Silviano Camberos S., a physician from Mexico, works with Kekchi Maya herbalist and bushmaster Jose Tot, on the left, to learn about local disease concepts and herbal treatments. Only after such discussions will he undertake ethnobotanical collections for pharmaceutical evaluation.

gories of disease, and the specialist vocabulary of his profession, you must work in *his* language, not yours," the anthropologist Bruce Biggs says, "because while you may, eventually, get to understand what he tells you in his language, and translate it into something that can be compared with Western ideas on the same topic, there is no way that your informant, perfect though his English may be, can do that for you. His very use of English will mask and obscure your topic of investigation." Formal instruction in an indigenous language is seldom available, so ethnobotanists assemble word lists, study published grammars, dictionaries, and translations of Western texts (often the Bible), and listen very carefully to indigenous speakers to learn their language.

Second, ethnobotanists must establish their own personal working ethnography, an understanding of the culture, of the group they are studying. The ethnographer James Spradley defines cultural knowledge as the "acquired knowledge people use to interpret and generate behavior." Some cultural knowledge, such as how to tie a knot or relate a legend, are explicit types of knowledge, knowledge that can be easily and quickly communicated to someone else. But many important types of cultural knowledge are tacit, outside the normal awareness of most members of the society: the amount of personal space that one maintains in interacting with other members of the culture; how to stand, sit, or position oneself in relation to others; when to speak and when not to speak; how loud to speak in the presence of village elders; and so forth. One can quickly and easily learn explicit cultural knowledge, such as modes of dress and types of food, by studying books or hearing the experiences of others, but tacit cultural knowledge can usually be obtained only by direct experience. Tacit cultural knowledge is much more important than explicit cultural knowledge in establishing rapport with healers. The best ethnobotanists are those who most rapidly learn and employ tacit cultural knowledge.

Most scientists formulate hypotheses, design research instruments, gather data, and then analyze the data, but this linear method is of little use to ethnobotanists in search of tacit cultural knowledge. They must work in an iterative, cyclical fashion, collecting tacit cultural information, employing it in their own behavior, interpreting the response to their efforts, and then refining their knowledge of cultural information before using it again. To do all this, ethnobotanists change their lifestyles to conform with that of the indigenous culture. Yet superficial changes in language, diet, and dress, while helpful, are seldom sufficient. Indigenous peoples are extraordinarily adept at sensing insincerity; it is our experience that genuine interest and clearly stated, respectful intentions help communicate to indigenous peoples the humility, trust, and respect that are so crucial for establishing rapport. Most experienced ethnobotanists are able to step for a time completely out of their own cultures and embrace the indigenous worldview as a new reality. If these efforts to establish rapport are successful, the ethnobotanist obtains a preliminary understanding of the culture's healing tradition.

Many earlier ethnobotanical studies, whether of healing techniques or other plant uses, simply produced a list of plants deemed "useful" by the people of an area. The ethnobotanist often made little effort to understand how the indigenous people viewed the plants in their own culture. Although these older-style surveys contain much useful information, particularly since we still know so little about the world's plant diversity, they are now in need of being repeated using newer techniques. These techniques have proven their value in studies of all kinds of plant uses.

Most commonly, the modern ethnobotanist adopts the role of participant-observer, living with the people under study, observing their daily life and customs, and learning about their lifestyle, foods, disease systems, and myths and legends. In true participatory ethnobotany, the indigenous person becomes a teacher, a colleague, and a respected and valued friend. These close relationships are not without some liability, however. Ethnobotanists may have trouble maintaining objectivity. Even more serious, formal interview techniques, which are designed to prevent an investigator from unconsciously directing the flow

and nature of the conversation ("leading" an interviewee), are difficult to maintain during participant observation. Because ethnobotanists consciously reduce the formal distance between observer and subject, they are vulnerable to the criticism that they move too deeply into indigenous paradigms. It should not be a surprise that many ethnobotanists become passionate advocates for indigenous rights, playing major roles in establishing indigenous-controlled reserves and ensuring that indigenous peoples share in the benefits of new discoveries such as medicines derived from plants.

The practitioners of another style of ethnobotanical research ask their indigenous colleagues to re-create events, perhaps those that were once more common. Ethnobotanists may ask to see how a limb is splinted with a palm leaf or commission aged shipwrights to build seacraft under a researcher's watchful

Samoan healer Lemau Seumanutafa and her apprentice instruct ethnobotanist Paul Cox (left) and pharmacognosist Lars Bohlin (right) on different types of medicinal plants.

eye. Although the information gained is valuable, the situations are by definition contrived: the patient is not actually in pain; a cross-oceanic migration is not imminent. The investigator is thus able to take notes in great detail. As more ethnobotanical research is carried out among Westernized peoples, the re-creation of past events becomes of greater importance.

Applying the "artifact/interview" method, pioneered by Brian Boom of The New York Botanical Garden, the scientist queries local people about an item constructed from plants. The investigator gathers information on where plants used to produce the object come from, then makes a trip to collect samples of the species used. Or the investigator may simply discuss the use of plants for food, medicine, or magic, without an artifact being presented.

Immersion ethnobotany, another new approach, reduces the distance between subject and observer still further, for the ethnobotanist using this method submits to being treated by an indigenous practitioner. For example, while in India studying the traditional medicinal system known as Ayurveda, Michael Balick was interned in a local hospital and given traditional Ayurvedic treatment by four practitioners, who applied a series of herbal massages and chiropractic manipulation, and also prescribed internal medicines. As a patient in treatment, Balick had, in effect, turned over control of the study to the traditional healers, in order to experience first-hand the very profound effects of the system in such a way that it could be described in detail.

Indigenous healing systems are often complex, but they incorporate at least three basic components: (1) a cosmological view of the universe that can help explain the cause, diagnosis, and treatment of disease; (2) a cultural context within which health care is given; and (3) a repertoire of pharmaceutical substances. We are unaware of any culture that does not possess such an indigenous pharmacopoeia. To keep track of an unfamiliar and often complex belief system, ethnobotanists document interviews with healers with copious notes, audio recordings, videotapes, and film. As they learn from the healers, they begin to see plants through the healers' eyes. Only then can they begin to accumulate a contextually significant collection of plants.

Specialist Healers in Belize

Many traditional healers in Belize are generalists, yet they have developed areas of specialization. Michael Balick and his colleague Rosita Arvigo have studied extensively with Hortense Robinson, a specialist in midwifery and other health

Left: Hortense Robinson, a traditional healer from Belize specializing in midwifery, prepares a poultice from species in the genera *Bursera* and *Hibiscus* to be used in the treatment of headache. Right: Andrew Ramcharan, a traditional healer specializing in the
treatment of snakebite, collects roadside plants in northern Belize. A surprisingly large
number of plants along roadsides and in secondary forests are valued for their medicinal
properties in Central America, as well as elsewhere in the tropics.

care issues involving women and children. She uses a specific set of plant
species, including some powerful species she considers too toxic for use by non-
specialists. One plant used by midwives is the castor oil bush (*Ricinus communis*
[Euphorbiaceae]). A leaf of this plant placed on a lactating mother's nipple is
said to reduce or stop the flow of milk. Other plants are used to treat irregular
menstruation or heavy menstrual bleeding.

Andrew Ramcharan, of Ranchito, a village in the north of Belize, specializes
in the treatment of snakebite in an area where the agroecosystem consists primarily of sugar cane and harbors many venomous snakes. His family came to
Belize from India. His grandfather, a well-known snakebite healer in Calcutta,
had learned his skills from his father, and they served him well in Central

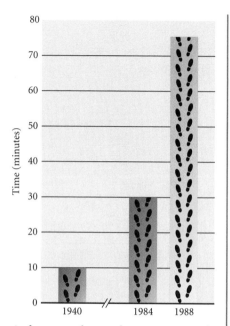

As forests are destroyed, so are vast repositories of plants used in traditional medicine. This graph shows the time required for a Belizean healer, Don Elijio Panti, to reach the secondary forest sites where he collects medicinal plants. Whereas in 1940 these sites were an average of 10 minutes from his house, by 1988 Don Elijio had to walk 75 minutes to reach an adequate site.

America. The Ramcharans found only some of the therapeutic plants they knew in Belize and experimented until they identified local species that could substitute for the taxa they had used in India. They also exchanged information with the local Maya Indians on plants they used. In such a manner, traditional medical systems evolve and adapt to local environments.

Both Hortense Robinson and Andrew Ramcharan provide primary health care services to the residents of their villages and surrounding areas, and, like many other traditional health care providers in Belize, they depend largely on medicinal plants that grow in the forest. As these forests are destroyed, increasingly greater levels of energy must be expended to locate and collect the plant medicines the healers use. Hortense Robinson often walks nine miles from her home to a remnant of a once-vast forest to collect plants she commonly uses in her practice. This sacrifice, she says, strengthens her spirit and ability as a healer, and increases her patients' faith in her work.

In addition to the loss of potential drugs far into the future—a loss that affects Western medicine as well as the indigenous population—an immediate consequence of deforestation is the degradation of the traditional primary health care system in the area. For many conditions, traditional medicine is effective, while also being less expensive, more widely available, and more culturally acceptable than Western medicine. Besides, traditional healers in Belize recognize and treat such diseases as *susto, viento,* and *envidio*—fright, wind, and envy. Western medicine cannot replace the medicine these healers practice.

Understanding these indigenous healing systems is not easy. Michael Balick, for example, asked people in Belize about plants that might be useful in treating cancer. He then collected individual species and provided bulk samples to a pharmacological laboratory for screening. Surprisingly, no more biological activity was observed in these plants than in another group that had been collected at random. Only later did Balick realize that "cancer" in Belize means a condition characterized by weeping, open wounds that are chronic, spreading, and difficult to heal—not what he had intended to research at all.

The Collection of Herbarium Specimens

Ethnobotanists are careful to document all the plants they collect with well-prepared voucher specimens that they deposit in herbaria. The importance of adequate herbarium voucher specimens cannot be overstated: if any question or

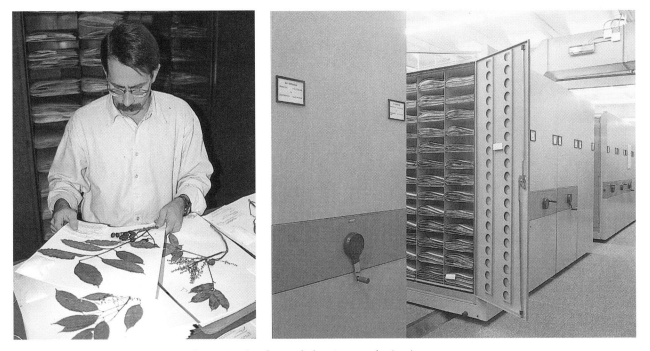

Left: Douglas Daly, a botanist specializing in the flora of the Amazon basin, is shown here identifying herbarium specimens of plants in the Frankincense family [Burseraceae]. The specimens serve as vouchers for his ethnobotanical and floristic studies in the extractive reserves of Brazil. Right: Cases in the herbarium of The New York Botanical Garden hold preserved plant specimens. Modern, climate-controlled storage facilities ensure that dried plant vouchers can be kept in collections indefinitely, for study by later generations of botanical scientists.

dispute should arise concerning the identity of the species involved, examination of a properly collected voucher specimen can unequivocally settle the matter. A properly documented voucher specimen should supply the ethnobotanical information, the names of the people interviewed, and a detailed description of the place where the plant was found so that the plant population can be located again if necessary.

Because of their scientific value for years to come (sophisticated bioassays of the future may require only micrograms of plant material), voucher specimens should be deposited and preserved in a well-curated herbarium and duplicate specimens should be deposited in geographically distant herbaria, including

Preparation of Herbarium Specimens

Preparation of an herbarium specimen begins with the selection of a plant representative of the population. The ethnobotanist collects all parts of the plant—the leaf, fruit, flower, and every other part a botanist would require to identify it. Plants in flower are not necessarily in fruit, and vice versa. For a good portion of the year, many plants are found in a sterile condition, especially those in tropical locations where the growing season is year round, or those in dormancy in the temperate region. And without a flower or fruit—what botanists call the "fertile" portion of the plant—it is often difficult to identify the species. If the plant is small, such as an herb, the entire organism can be collected and pressed for preservation.

The photo at the bottom of this page shows a poor herbarium specimen. It is described as coming from a plant 2 meters tall in the primary forest and as having a yellow flower. Unfortunately, the flower has not been collected, and only leaf fragments are present on the herbarium sheet. One specialist who examined it could not place it in the proper family; another could say only that it was definitely not a palm. The species is used in Ecuador to thatch roofs and tie bundles, but as far as the ethnobotanist is concerned, the specimen is unidentifiable.

The photo of the herbarium specimen at the top of this page shows a medicinal plant used to treat skin burns. It is identified as a vine that grows to 2 meters tall and has purple tubular flowers and green fruits that turn red at maturity. Note that both fruits and flowers are present on the specimen, as well as a good section of the vine and numerous leaves. This plant can easily be

Top: An example of a well-prepared herbarium specimen, containing fruits and flowers, that can be identified as *Lycianthes lenta*, in the tomato family [Solanaceae]. Bottom: An example of a poorly prepared herbarium specimen. The collector has gathered only a few fragments of the leaf, with no flowering or fruiting material. Since it is impossible to identify this plant from the material present, making the ethnobotanical research that is based on this specimen is of little value.

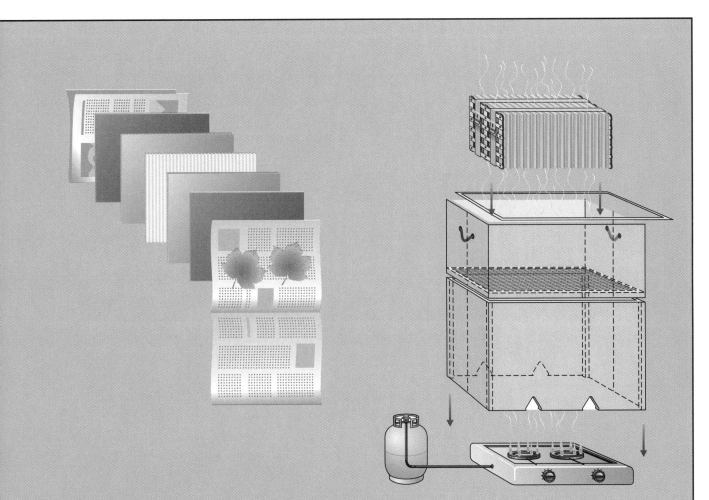

Herbarium specimens are produced by pressing and drying plants that have been collected in the field. A plant press, left, is a standard tool. A plant sample, typically of leaves, flower, and fruit, is placed flat inside a sheet of folded newspaper, then pressed flat by placing it within a sandwich of blotter paper (red), cardboard, and corrugated aluminum. Many such sandwiches are placed between wooden endboards that are tightly clasped by buckled straps. The entire press is placed in a plant drier, as shown on the right. This dryer consists of a box with a screened bottom to prevent leaf and paper fragments from falling into the flames; the box rests on a larger wooden frame with slots or short legs that allow air to enter at the bottom—there the air is heated by a propane or kerosene burner or by lightbulbs or a small heater if electricity is available. The warm, dry air rises upward and moves through the plant press via holes in the corrugates, allowing moisture to be removed from the blotters and plant specimens. Most plants, except those with fleshy parts, can be dried in 6 to 24 hours using this apparatus.

(box continued on next page)

identified as belonging to the family Solanaceae, specifically as *Lycianthes lenta*. Both of these plant specimens were made to document an ethnobotanical use by indigenous people, but only the one on the top of page 48 can be properly identified.

It is important to accompany each specimen with careful notes that tell where it was collected—the latitude and longitude, the village, the state or county, and the country. The notes should also provide as complete a botanical description as possible—the size and shape of the plant, the colors of flowers and fruits, any fragrance—especially if the plant is too large to be completely preserved on the herbarium sheet. A palm, for instance, may be 30 meters tall; the leaves alone can measure 8 meters. In this case, representative pieces of flower, fruit, leaves, and stem sections can be taken. The notes should also describe how the plant was collected, and the process is often documented by photographs. The date when the plant was collected and the names of all members of the collection team should be provided in the lower portion of the label. Finally, the contributions of institutions and foundations that have supported the research must be acknowledged.

Usually plants are pressed in sheets of folded newspaper and preserved for a time in an alcohol bath if they cannot be dried in the field. An alcohol bath (usually less than 50 percent alcohol) compromises the chemical integrity of the herbarium specimen, but if a botanist cannot manage to carry a heater, sheets of cardboard and corrugated paper, ventilators, wooden presses, and straps into the area, immersing the plant in a fluid preservative until it can be dried is the best thing to do.

Once the material is delivered to the institution where it is to be studied, the plants are sewn or glued to high-quality acid-free rag bond paper and stored in steel cases in an environmentally controlled area—an herbarium. Under proper storage conditions, herbarium specimens will retain their scientific value almost indefinitely.

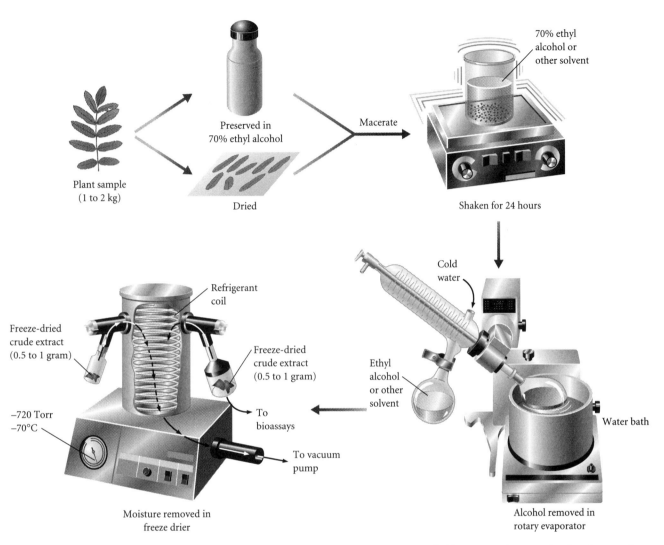

Plant sample
(1 to 2 kg)

Preserved in
70% ethyl alcohol

Dried

Macerate

70% ethyl
alcohol or
other solvent

Shaken for 24 hours

Cold
water

Refrigerant
coil

Freeze-dried
crude extract
(0.5 to 1 gram)

Freeze-dried
crude extract
(0.5 to 1 gram)

Ethyl
alcohol
or other
solvent

To
bioassays

–720 Torr
–70°C

To vacuum
pump

Water bath

Moisture removed in
freeze drier

Alcohol removed in
rotary evaporator

ones in the countries where the material was collected. Ethnobotanical data on the diseases treated, the mode of formulation, and the methods of administration should be copious in order to guide later investigations. The collection number of the specimen should be used to label all subsequent pharmacological fractions and residues so that any new discovery or question can be immediately referred to the original herbarium sheet.

Ethnobotanists are responsible for preparing not only voucher specimens but also materials for pharmacological testing. They must note carefully

To prepare plant samples for pharmacological testing, a botanist first collects 1 to 2 kilograms of plant parts in the field and preserves them by drying or by placing them in a fluid preservative. In the laboratory, the plant materials are macerated, placed in a solvent, and shaken for 24 hours. The solvents are removed in a rotary evaporator, and the plant extract is then freeze-dried. The process yields 0.5 to 1 gram of crude extract, which can be tested in various bioassays.

the parts of the plants the healers use, since flowers, leaves, shoots, and roots often differ significantly in their chemical composition. Dried samples ranging from 0.5 to 2 kilograms have traditionally been supplied for testing, but drying may not always be the best method of preservation, for heat can destroy some classes of chemicals. Storage in alcohol or freezing are splendid ways to preserve plant materials, but in many parts of the world it is not very practical. In general, the best ways to collect and preserve plants are the ways the healers use.

The ethnobotanist collects an initial sample of approximately 0.5 kilogram of plant material along with an herbarium specimen and takes them to the laboratory. There fractions of the sample containing different components are extracted with a variety of aqueous and organic solvents. The extracts are then tested against various bioassays to identify promising pharmacological leads and novel activity.

Bioassay procedures have evolved from recording observations of live animals that have been dosed with plant extracts, to high-volume, sophisticated *in vitro* procedures that determine if the plant extract inhibits specific enzymes, binds to certain molecular receptors, or exhibits other types of highly specific biologic activity. If the extract shows bioactivity, the botanical team is then directed to return to the area where they first collected the plant to retrieve a bulk sample, often 50 to 100 kilograms of material. At that time the botanists compare the new specimens with the original material and verify the botanical identification. Sufficient plant material is gathered to permit fractionation and structural elucidation of the chemical components responsible for the identified activity. Scientists compare the chemical structure with known structures to see if the entity has been previously discovered. After they have isolated the chemicals and determined their structure, they decide whether or not to attempt to synthesize the compound. Various factors—cost, the quantity needed, the availability of the natural resource—impact this decision.

The compound is then entered into clinical trials in which its effects on human subjects are observed. In Phase I clinical trials, researchers watch for any toxicity of the compound when it is administered to human volunteers. In Phase II trials, they determine the efficacy of the compound against the disease in a small group of people. In Phase III trials, they study a much larger population of patients under rigorous clinical conditions. After the efficacy and safety of the substance have been demonstrated, a new drug application (NDA) can be submitted to the Food and Drug Administration for the use of the compound as a therapy in the United States.

Examples of Samoan Disease Terminology

Anufe	Intestinal worms
Ate fefete	Swollen liver
Malaga umete	Head ulcerations
Failele gau	Complications of maternity
Fe'efe'e	Untranslatable internal disease
Lepela	Leprosy
Lafa	Ringworm

Samoan disease names are derived in a variety of ways. *Anufe,* for example, means "worms"; the disease takes its name from the cause. *Ate* (liver) and *fefete* (swollen) provide an anatomical origin for a disease name, while *malaga umete* (*umete* = bowl) refers to the shape of the ulceration. New mothers are vulnerable to *failele gau,* and *fe'efe'e* (octopus) summons a vision of tentacles crawling inside one's intestines. *Lepela* is a transliteration of the word "leprosy." *Lafa* is an irreducible term for ringworm.

Isolation of the Anti-HIV Drug from a Samoan Tree

Except for a group of basic remedies that nearly all Samoans know, herbalism is a specialty practiced by healers called *taulasea*—herbalists—nearly all of whom are women. They have learned their craft from their mothers or other female relatives. Some Samoan *taulasea* use more than 100 species of flowering plants and ferns. The number of Samoan herbalists has dwindled. Those still practicing are very old and few have apprentices.

Samoan medicine differs significantly from Western medicine in its descriptions of disease etiology. As in Belize, many diseases recognized in Samoa are not directly translatable into Western terminology.

Samoan healers refuse payment for their services, arguing that the plants are a gift of God. Yet their knowledge is formidable: a typical healer can identify

The late Samoan healer Mariana Lilo prepares a tea used to treat hepatitis from *Homalanthus nutans.* To produce the remedy, she immerses the macerated bark of *H. nutans* in boiling water. Her patient will ingest only the tea containing the water-soluble fraction of the bark.

over 200 species of plants by name, recognize over 180 disease categories, and compound more than 100 remedies.

Samoan healers may treat ailments with massage, special diets, or incantations. When a healer diagnoses a disease that requires an herbal treatment, she immediately begins to collect the necessary plant materials, since only fresh plants are used. Most Samoan remedies are formulated from flowering plants. Formulation techniques are specific to the plant part used. Most remedies are water infusions, some are oil infusions, and a few are ignited and inhaled. Many remedies, including those for internal ailments, are applied externally. Most treatments are prepared using a combination of several species of plants gathered in the wild, but some remedies are prepared from a single species.

Consider the treatment that Epenesa Mauigoa uses for *fiva samasama* (*fiva* = fever; *samasama* = yellow), the clinical manifestation of acute hepatitis. After confirming a diagnosis of *fiva samasama,* Mauigoa has one of her children journey to the forest for the wood of the *mamala* tree (*Homalanthus nutans* [Euphorbiaceae]). But not any type of *mamala* will do; botanists recognize one species of *Homalanthus,* but Mauigoa recognizes two. "Only the *mamala* with long white petioles is used," she explains. The *mamala* with red petioles on the leaf is reserved for abdominal complaints called *tulita saua*.

After her daughter returns with *Homalanthus* wood, Mauigoa scrapes away the outer cork and epidermal tissues of the wood and extracts the "inner bark," or cambial tissue, by scraping it with a knife. She places the scrapings in a cloth, ties it like a tea bag, and immerses the bag in boiling water for half an hour. After removing the bag and discarding its contents, she filters the liquid through a cloth and gives it to the patient to drink.

In 1984 Epenesa Mauigoa and other healers told Paul Cox about this remedy for *fiva samasama*. Gordon Cragg of the Natural Products Branch of the National Cancer Institute had agreed to evaluate the pharmacological efficacy of medicinal plants that Cox might find in Samoa, and among the materials he collected were stem wood samples of *H. nutans*. Attempting to simulate traditional preparation techniques, he chose not to use the standard method of air drying to prepare samples for analysis, but instead returned to his laboratory in the United States with the samples preserved in aqueous alcohol in aluminum bottles. In his lab he removed the alcohol in a rotary evaporator and placed the extracts in a freeze drier. Cox then carried the freeze-dried samples to the NCI in Maryland, where a team including Michael Boyd, John Cardellina, Kirk Gustaffson, Peter Blumberg, John Beutler, and other researchers tested them for activity against the HIV-1 virus—the virus associated with acquired immunodeficiency syndrome (AIDS).

The NCI team soon found that the stem wood extracts exhibited potent *in vitro* activity against the HIV-1 virus, both stopping the virus from infecting healthy cells and preventing infected human cells from dying. Bioassay-guided fractionation resulted in the isolation of prostratin (12-deoxyphorbol 13-acetate).

The identification of prostratin, which belongs to a group of compounds known as phorbols, as the active component in *H. nutans* caused some concern: phorbols are known tumor promoters. Research conducted by a team led by Peter Blumberg at NCI demonstrated that prostratin does not promote tumors, even though it activates protein kinase C, a typical indicator of tumor promotion. Indeed, the NCI team found that prostratin functions as an antipromoter: it stops mutant cells from developing into tumors. The NCI is currently soliciting bids from drug companies to license prostratin for drug development. Since prostratin stops cells from becoming infected with the HIV-1 virus and prolongs the life of infected cells, it may prove to be effective as part of a combination therapy in tandem with proteases and other antiviral compounds. Yet toxi-

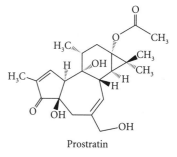

Prostratin

The National Cancer Institute (NCI) has declared prostratin, a molecule derived from the Samoan tree *Homalanthus nutans* to be a drug candidate for the treatment of AIDS. The NCI is accepting bids from pharmaceutical firms to license and develop this drug.

city may be a problem. Although prostratin has given no signs of promoting tumors, as a phorbol it is a member of a very toxic group of chemicals. Only careful toxicological studies will determine if prostratin can be safely advanced to human clinical trials.

Ethnobotany and the Future Discovery of Drugs

How many and what types of drugs remain to be discovered through the ethnobotanical approach? Is there any way to estimate the probable success of the ethnobotanical approach in the future?

Some estimate of the prospects can be gleaned from an analysis of maladies for which healers administer treatments prepared from plants. A review of published accounts of plant uses in 15 widespread geographical areas—Australia, Fiji, Haiti, India, Kenya, Mexico, Nepal, Nicaragua, North America, Peru, Rotuma, Saudi Arabia, Thailand, Tonga, and West Africa—makes it possible to categorize the plants according to the ailments for which they are used. The categories found include diseases of the nervous and cardiovascular systems, obstetrical and gynecological ailments, treatment of neoplasms (cancer), gastrointestinal ailments, skin diseases, inflammation (including fevers), microbial diseases, renal ailments, hydration therapy, parasitic diseases, immunotherapy, blood diseases, and poisons. We can compare these indigenous uses of plants with the Western drug uses reflected in the *United States Pharmacopoeia*. A similar approach allows us to categorize the 50 ethnobotanically derived drugs listed on pages 34–35.

Such an analysis shows a striking difference between the mean percentages of disorders treated with indigenous plants and Western drugs. Indigenous plant remedies are focused more on gastrointestinal (GI) complaints, inflammation, skin ailments, and ob/gyn disorders, whereas Western drugs are more often used to treat disorders of the cardiovascular and nervous systems, neoplasms, and microbial ailments. Why these differences? There are several possible answers:

1. *Perceived peril.* Cardiovascular illness, neoplasms, microbial infections, and nervous system ailments are the biggest killers in Western cultures. Indigenous peoples, who do not have the lifestyles or predicted life spans associated with cardiovascular disease and cancer, see diarrhea, complications of maternity, and inflammation as more perilous.

2. *Saliency.* Indigenous peoples can easily detect inflammation, skin diseases, and GI ailments, but most cancers and cardiovascular disease are difficult

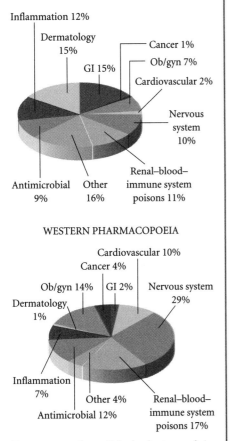

Percentages of medicinal plants used in various treatment categories by indigenous peoples of 15 countries compared to the percentages of drugs used in the same treatment categories by Western societies. Note how indigenous treatments focus on dermatology, inflammation, and gastrointestinal ailments, while Western drugs are more likely to be used for cancer, heart ailments, and antimicrobial remedies.

to diagnose by traditional methods. In fact, few indigenous languages have a word for cancer, leukemia, lymphoma, or hypertension.

3. *Toxicity.* Indigenous peoples are likely to avoid plant medicines that are highly toxic in low doses. Most cardiovascular and anticancer drugs, as well as those that act on the central nervous system, have extremely narrow dosage windows and thus are not likely to be acceptable to indigenous peoples. (Withering's dosage problem with digitalis demonstrates this problem.)

4. *Economic incentives.* The discovery of drugs in the Western world is driven by market considerations. When we analyze the amount of money spent in the United States for research in the various treatment categories, we find that cardiovascular illness, neoplasms, nervous system disorders, and microbial diseases receive 72 percent of every research dollar. Such economic pressures are unique to Western scientists; they do not affect indigenous healers. Thus the percentages of drug types discovered by the ethnobotanical approach more closely represent funding opportunities than indigenous use categories.

On the basis of this analysis, we can predict success for properly designed ethnobotanical surveys for gastrointestinal, anti-inflammatory, ob/gyn, and dermatological drugs. But does this mean that no new cardiovascular, anticancer, or antimicrobial drugs are likely to be found by the ethnobotanical method? Are new anticancer drugs such as vincristine and new cardiac drugs such as digitalis still waiting to be discovered?

An analysis of the table on pages 34–35 suggests that new drugs in these categories indeed await discovery by the ethnobotanical approach. Of the 50 drugs listed, 22 percent are cardiovascular substances (compared to 2 percent of indigenous plant remedies), 20 percent are used for the nervous system (compared to 10 percent of indigenous plant remedies), and 10 percent are used for neoplasms (compared to only 1 percent of indigenous plant remedies). Ethnobotanists' success in finding drugs at a rate far higher than predicted bodes well in general for this approach to drug discovery. "Seek and ye shall find" seems to be the operative principle.

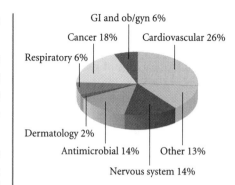

Percent of each research dollar spent for drug development in the United States, by disease categories.

Ethnobotanical Research and Traditional Health Care in Developing Countries

So far we have focused largely on ethnobotanical research in efforts to discover new drugs for Western medicine. Yet according to recent estimates by the World Health Organization, more than 3.5 billion people in the developing world rely

Don Elijio Panti, the late Maya traditional healer from Belize, is shown treating one of the thousands of patients that come to him each year. A vital part of their country's health care system, specialists in traditional medicine provide health care to a large portion of the world's population. Unfortunately, there are few in the younger generation who are being trained to take the place of people such as Don Elijio.

on plants as components of their primary health care. Just as many Europeans know of the use of *Aloe vera* [Aloaceae] to treat burns, many indigenous peoples know of some common plants that have medicinal uses. Ethnobotanical research should not be limited to discovering new pharmaceuticals for Westerners; it can also be of some benefit to peoples in developing countries.

An increasing number of nations, including China, Mexico, Nigeria, and Thailand, have decided to integrate traditional medicine into their primary health care systems. In these systems, ethnobotanical research plays a crucial role in documenting the traditional health care practices of the country. Medicinal plant lore often recedes or completely vanishes in the wake of rapid Westernization. In some countries, careful ethnobotanical studies have become invaluable records of ancestral ways. In areas where the people are moving away from traditional lifestyles, particularly in rapidly growing urban populations, careful ethnobotanical documentation can provide the needed foundation for educational programs. Workers at Mahidol University in Bangkok, for example, have prepared a series of slide presentations and pamphlets to teach schoolchildren about traditional Thai uses of plants.

Ethnobotanical research can also help in the discovery of crude drugs. Only pure compounds with known structures and pharmacological activities are permissible as drugs in Western medicine, but in many developing countries the price of such pure substances puts them beyond the reach of all but the affluent. Careful clinical studies can document the safety and efficacy of crude extracts or tinctures of plants that can be dispensed at far less cost. Carefully designed clinical trials of crude botanical drugs have been conducted in Mexico and Thailand. The trials in Thailand have resulted in certification of a tincture of the beach morning glory, *Ipomoea pes-caprae* [Convolvulaceae], as an anti-inflammatory treatment.

An area of ethnobotanical drug discovery that has yet to be developed is that of "gray pharmaceuticals"—drugs of proven safety and efficacy that are not marketable in the Western world. Decisions concerning marketability in the Western pharmaceutical industry are not driven solely by proof of safety and efficacy. To be marketable, a drug candidate must affect only one point on a biochemical pathway: compounds that affect multiple points of the same pathway are unlikely to be marketed because only "magic bullets" (single-activity drugs) are viable in today's legal and economic environment. Drug candidates must also show superiority over competing drugs in the same market. Thus some plant-derived drugs that are not marketable as Western pharmaceuticals may still be acceptable in the country of their origin, particularly if they can be

Throughout the South Pacific and Southeast Asia, an extract of the beach morning glory, *Ipomoea pes-caprae,* is used on the skin to treat inflammation. In Thailand, a tincture of this plant is now sold in drugstores after first being proved safe and effective in rigorous clinical trials.

produced cheaply. The transfer of information (sometimes costing millions of dollars) concerning the safety and efficacy of such gray pharmaceuticals from Western firms to developing countries, along with the appropriate patent rights and technologies to enable the developing countries to produce them, should be encouraged.

Safeguarding Indigenous Intellectual Property Rights

We do not know what compensation, if any, Withering offered the old woman in Shropshire who guided him in his discovery of digitalis. The Aymará Indian, Manuel Incra, who collected the seeds of the quinine-rich *Cinchona ledgeriana* in Bolivia for Charles Ledger paid for his generosity with his life. Such treatment of indigenous peoples was not unusual. Historically, the intellectual property rights of indigenous peoples have not been recognized. The use of information supplied by indigenous peoples in the discovery of commercially marketable pharmaceuticals raises the question of those people's intellectual property rights and the ownership of biodiversity.

The indigenous healers we work with offer significant intellectual guidance and input into our research programs. Thus we prefer to call them "colleagues," "guides," or "teachers" rather than "informants," the term favored by anthropologists. In view of their significant intellectual contributions to our research, we believe that indigenous peoples are entitled to the same intellectual property rights enjoyed by other investigators. In the case of prostratin, for example, the National Cancer Institute and Brigham Young University have guaranteed that a significant portion of any royalty income will go to the Samoan people.

Yet in many cultures, the preservation of important habitats is equally urgent. In Samoa, four village-owned and -managed reserves totaling 50,000 acres, beginning with the Falealupo Rain Forest Reserve (where the tree that produces prostratin was first collected), have been created with donated funds. And in Belize, the world's first extractive reserve for medicinal plants has been created on 6000 acres of tropical rain forest by the local government working with the association of traditional healers with significant international support. This effort seeks to demonstrate that conservation and the use of forests as sources of locally consumed medicines are compatible objectives. Elsewhere, as in India, medicinal plant reserves are being established to ensure a continued supply of plants for traditional health care practitioners and their patients.

Cash disbursement of royalty income most closely approaches the Western concept of equity, but this approach fails with peoples who have no monetary system. For many indigenous peoples, the right to live unmolested and undisturbed on their ancestral lands is the greatest value. Establishment of nature preserves that protect both biodiversity and indigenous cultures is of tremendous importance to indigenous peoples. And this need can be most clearly seen in those societies that depend neither on commerce nor on agriculture for their sustenance: the hunter-gatherers.

A vendor offers fresh ginger rhizomes in an open air vegetable market in Pune, India. Markets are exciting places to study the diversity and local uses of plants, especially those employed as foods and medicines.

From Hunting and Gathering to Haute Cuisine

I magine life in the Kalahari. This desert, a vast arid plain of dried rivers and salt pans, covers 100,000 square miles in the southern part of the African continent. Parts of the Kalahari have fewer than 6 inches of annual rainfall, and even that little bit of precipitation is erratic and unpredictable. Daytime temperatures can average 100°F (38°C). Severe and prolonged droughts are common. Soils are sandy and salty and completely unsuitable for cultivation. As a result, the Kalahari is one of the world's most sparsely populated regions. In its lack of arable soil, low precipitation, and temperature extremes the Kalahari is similar to another inhospitable region: the treeless tundra of Alaska. Yet both the Kalahari Desert and the Arctic tundra have been successfully inhabited for thousands of years—not by agriculturalists, but by people who have depended on the wild plants they gather (in the Kalahari) and on the seals, walruses, and whales they hunt (in the Arctic). Because they rely on these wild sources of food, the 50,000 San Bushmen of the Kalahari and the 28,000 Inuit of Alaska and the Aleutian Islands are called hunter-gatherers.

"Hunter-gatherers" is the term applied to those people that exist by gathering wild plants, fishing, hunting, and foraging for invertebrates. The few cultures that continue to rely solely on hunting or gathering for subsistence represent extremes on a cultural spectrum of mixtures of agriculture and wild harvesting. Most societies that focus primarily on hunting dabble in agriculture, and most agrarian societies occasionally hunt and gather wild plants. Some indigenous cultures represent transitional states between subsistence foraging and agriculture. The King's College anthropologist David Harris has found, for example, that the natives of the Torres Strait Islands, south of Papua New Guinea, weed and tend the wild cycads from which they gather seeds—activities that resemble a precursor to true agriculture. And many agrarian societies use wild plants and the meat of animals they hunt to supplement agricultural produce.

Agrarian societies have generally considered hunter-gatherer societies to be primitive, but in reality hunter-gatherer cultures are characterized by close family ties, an abundance of leisure time, and a remarkably sophisticated knowledge of indigenous plants. For these reasons the University of Chicago anthropologist Marshal Sahlins terms the San Bushmen as "the original affluent society." It now appears that hunter-gatherer cultures are markedly superior to agrarian societies in their resiliency in the face of environmental perturbation, their low level of organized warfare, and their reliance on a relatively large number of plant species.

If we accept hunting and gathering as a preagricultural state, then it appears that the rise of agriculture has been accompanied by increased dependence on a reduced number of staples. Hunter-gatherer societies had no such dependence on cultivated staples and were able quickly to shift patterns of caloric intake within a large palette of wild-gathered species. Though many agrarian societies engaged in hunting, reef foraging, and wild plant harvesting, these activities were unable to take the place of agriculture during times of famine.

Just as hunting and gathering gave way to an agricultural way of life in most places, the traditional ways of present-day agricultural societies are changing as more and more indigenous peoples confront modernity. And just as the early farmers lost some of the benefits of hunting and gathering, contemporary indigenous societies are losing benefits that had been provided by traditional diets, which, ethnobotanists and others are discovering, are sometimes remarkably adapted to people's needs.

Diabetes and Traditional Diets

Consider the Akimel O'odham tribe (also known as the River Pima tribe) of the Sonora Desert of Arizona. Some anthropologists consider the Akimel O'odham to be descendants of the prehistoric Hohokam culture. They traditionally lived by harvesting wild plants and cultivating beans and corn. The tribe continued relatively unmolested until the mid–nineteenth century, when white settlers began to divert water from the Gila River. By 1870, tribal life was disintegrating. Today the Akimel O'odham live on reservations near Phoenix, Arizona. They buy most of their food in stores, and such Western staples as white bread and rice figure prominently in their diet. Unlike other North Americans, however, the Akimel O'odham have a high rate of diabetes—in fact, the highest in the world: more than 50 percent of the adult population are afflicted by the disease.

The ethnobotanist Gary Nabhan at the Desert Botanical Garden near Tucson began to wonder if this high rate of diabetes was related to the change from the Akimel O'odham's traditional diet to a Western diet. For centuries the Akimel O'odham existed on a diet of wild and cultivated legumes, cactus pads and fruits, corn, mesquite pods, and acorns. Today their diet has shifted to

Prickly pear cactus (shown in foreground) and other native plants of the Sonoron desert once composed the primary diet of the Akimel O'odhem, or River Pima, tribe.

processed wheat products, sugar, coffee, and cereals common to modern North American diets. Nabhan decided to test whether this dietary shift may have exacerbated diabetes in the Akimel O'odham by examining how their traditional foods affect blood sugar and insulin response.

Nabhan and his co-workers Janette Brand, Janelle Snow, and Stewart Truswell had volunteers eat cakes made from mesquite pods (*Prosopis velutina* [Mimosaceae]), a broth made from tepary beans (*Phaseolus acutifolius* [Fabaceae]) and lima beans *(Phaseolus lunatus)*, a stew of venison and acorns (*Quercus emoryi* [Fagaceae]), and hominy made from traditional Pima corn (*Zea mays* [Poaceae]). They then tested the blood sugar level and glycemic index of the volunteers, an index of how rapidly the body produces insulin to metabolize sugar. Blood sugar always rises after a meal, as sugars released in the course of digestion are absorbed into the bloodstream. In response, the pancreas releases the hormone insulin, whose function it is to signal fat and muscle cells to absorb the sugars and clear them from the bloodstream. High blood sugar levels for an extended period of time produce the damaging consequences of diabetes. Analysis showed that the slowly digested carbohydrates characteristic of the traditional Akimel O'odham diet resulted in a slow rise in blood sugar levels and a low insulin response. The glycemic index of even the Akimel O'odham corn, the highest of all the traditional foods tested, was lower than that produced by modern sweet corn. All of the traditional foods resulted in a lower insulin response than that of modern white bread. One conclusion seems inescapable: the high incidence of diabetes among the Akimel O'odham

The pancreas plays a key role in maintaining constant levels of blood sugar. When blood sugar is low (right), the pancreas secretes glucagon, which signals the liver to release glucose, thereby increasing blood sugar levels. When blood sugar is high (left), the pancreas secretes insulin, which signals cells to increase their absorption of glucose, thereby lowering blood sugar levels. Impairment of insulin secretion can lead to diabetes.

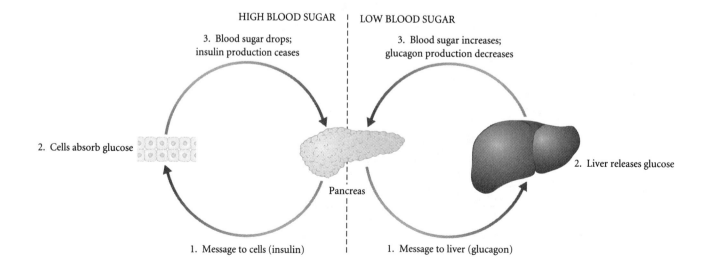

HIGH BLOOD SUGAR | LOW BLOOD SUGAR

3. Blood sugar drops; insulin production ceases

3. Blood sugar increases; glucagon production decreases

2. Cells absorb glucose

2. Liver releases glucose

Pancreas

1. Message to cells (insulin) | 1. Message to liver (glucagon)

is probably a consequence of their newly adopted Western diet, increased over-all caloric intake, and the health risks inherent in a sedentary lifestyle.

Why do traditional River Pima foods result in a lower insulin response? In part, because the starches in mesquite pods, acorns, and beans are digested more slowly than the starches in Western foods. But there are other reasons as well: the increased fiber content of traditional foods slows their absorption in the small intestine, so the rise in blood sugar is more gradual. Mucilage present in mesquite pods and cactus pads also dramatically lowers the insulin response by slowing the digestion and absorption of starches. Even the traditional Pima ways of grinding and processing render the foods less likely to exacerbate diabetes.

Nabhan's findings are very similar to those of Anne Thorburn and her co-workers, who studied changes in the diets of Australian Aborigines. The bush potato, *Ipomoea costata* [Convolvulaceae], produces far lower glucose and in-sulin responses than regular potatoes, *Solanum tuberosum* [Solanaceae], in Aborigines, but, interestingly, not in Caucasians. If such differences in starch metabolism are heritable, they could suggest a specific genetic response to slow-release carbohydrates among Australian Aborigines. This might account for a high rate of diabetes among Aborigines who shift to a Western diet. However, such dietary influences are almost surely of lower significance than sedentary lifestyles, increased caloric consumption, and resultant obesity.

Evidence of a relationship between genetic inheritance and response to diet has been published in the *New England Journal of Medicine* by Jeremy Walston of Johns Hopkins University and his collaborators. Intrigued by evidence that non-insulin-dependent diabetes mellitus (NIDDM) is heritable, the team de-cided to investigate the molecular basis of the disease in a population where NIDDM is very high: the Akimel O'odham of Arizona. The receptor molecule of interest to Walston was one of the many types of receptor molecules whose job it is to detect the presence of a particular molecule and send a signal that affects energy expenditure and the metabolism of certain fat tissues. These in turn affect the basal metabolic rate—the rate at which the body burns energy when at rest. Walston reasoned that since a low resting metabolic rate is a risk factor for obesity, and since obesity is a major risk factor for diabetes, perhaps mutations in this gene could lead to diabetes.

Walston's group studied 642 Akimel O'odham and Tohone O'odham Indians between 35 and 87 years old. They measured each subject's resting metabolic rate, total body fat, blood glucose and insulin concentrations after oral glucose administration, and tested each subject for the presence or absence of diabetes. They studied the receptor gene in ten unrelated obese Indians with

NIDDM for possible mutations and found mutations in the gene in all ten of them. The mutant genes produced a protein that has an altered biological activity.

Walston and his collaborators then screened 390 Indians with NIDDM and 252 Indians without the condition for the altered protein. They found no statistically significant association between the altered protein and NIDDM in mature Indians, but they did find a correlation between the presence of the altered protein and early onset of diabetes. Furthermore, they found a strong relationship between the mutant protein and a low resting metabolic rate; Indians who have the mutation expend on the average 83 calories less per day than those who lack the mutation. Thus the mutation may not directly cause NIDDM, but the lowered metabolic rate it entails increases the likelihood of obesity, which accelerates the course of the disease. Perhaps this explains why the number of Akimel O'odham males over 45 who have inherited the mutation from both parents is lower than expected: such people are more likely to die young of NIDDM.

These research results have significance far beyond the Sonoran Desert. Walston linked obesity, a health problem that afflicts a third of all people in industrialized populations, to a possible genetic source. If we consider only the genes for this receptor, then 31 percent of the receptor genes in the Pima Indian population carry a mutation. This figure compares with 13 percent for Mexican Americans, 12 percent for African Americans, and 8 percent for white Americans. Identification of the gene will probably lead to better dietary counseling for those at risk, and it may also open the possibility of gene therapy for those with a genetic proclivity toward a low resting metabolic rate and resultant obesity.

Is the Akimel O'odham's traditional diet of mesquite pods, acorns, and beans a response to the mutant receptor genes prevalent in the population? Or has the mutation been maintained in the population because a lowered resting metabolic rate reduces the likelihood of starvation during a famine? It is tempting to hypothesize an interplay among genes, diet, and culture in the Akimel O'odham, but further work is necessary to test this conjecture.

It is rash, however, to assume that all unusual relationships among diet, health, and culture have a genetic source. As Richard Lewontin has pointed out, pellagra, which is caused by a vitamin-deficient diet, was once believed to have a genetic origin because it recurred from generation to generation in poor families. Genetic responses to indigenous diets should be vigorously tested rather than assumed. The danger of assuming a genetic link between diet and health has been demonstrated by studies of the Masai and Batemi cultures of East Africa.

High Meat Consumption, Low Cholesterol Levels, and Indigenous Plants: New Leads from Kenya

The Masai and Batemi are cattle-herding peoples of Kenya and Tanzania, numbering over half a million in the most recent census. In traditional settings, the Masai live almost exclusively on the meat, milk, and blood provided by their cattle, a nutritional regime that the *New Scientist* has called "the world's worst diet." Although the Batemi, as agropastoralists, do not depend on meat as heavily as the Masai, average members of these tribes ingest up to 2000 milligrams of cholesterol a day—well over the maximum recommended daily intake of 300 milligrams. Yet, remarkably, their blood cholesterol levels are low—about one-third the cholesterol level of the average American. Researchers have long argued that these peoples must have specialized genes that prevent them from succumbing to what otherwise appears to be a sure prescription for arteriosclerosis and resultant cardiac disease.

The supposition of a genetic response to a high-cholesterol diet does not, however, account for the high cholesterol levels found in Masai and Batemi who move to Nairobi and switch to a Western diet.

Timothy Johns, an ethnobotanist at McGill University, has studied a variety of wild-gathered tree barks that the Batemi and Masai people add to their meat. Typically they boil meat slowly in milk, stirring the bark of *Acacia goetzei* [Mimosaceae] and *Albizia anthelmintica* [Mimosaceae] and other trees into the

The Masai people of Africa, who depend on "the world's worst diet" of meat, blood, and milk, have amazingly low cholesterol levels, perhaps due to bioactive compounds in some of the tree bark they add to their stews. Here a Masai is preparing a stew that includes the bark from a species of *Acacia*.

broth. They do not use these barks as flavorings in other foods, so Johns began to suspect that the barks might play a more sophisticated role: perhaps they serve to lower blood cholesterol. With his colleague Laurie Chapman, Johns tested the barks and found evidence that they do reduce blood cholesterol, possibly because they contain unique saponins, organic compounds with a sugar-like core that form a soapy foam when they are dissolved in water. The implications of this research are profound: if new cholesterol-lowering substances can be isolated from these wild-gathered plants, millions of people could reduce their risk of cardiovascular illness, the gravest health threat in the Western world.

As the Akimel O'odham Indians, the Australian Aborigines, and the Masai and Batemi of East Africa have demonstrated, a Western diet poses a grave threat to more than Westerners. Throughout the world, indigenous peoples are increasing their consumption of Western foods. In some societies, such as Samoa, imported foods such as canned corned beef have acquired significant social prestige, but we now know that the high fat and sodium levels of Western diets are correlated with increased triglyceride levels, coronary disease, obesity, and diabetes.

Diets in Transition: Food as Medicine

Diet-related diseases are particularly striking among indigenous agrarian and pastoral societies that have only recently adopted Western ways. The Sami or Lappish people of Scandinavia have suffered from skyrocketing levels of coronary disease as they have relied more on purchased foods and less on their reindeer herds and wild-gathered nuts and berries. In Samoa, obesity, diabetes, and coronary disease have dramatically increased as the islanders have exchanged their traditional diet of root crops and fish for imported rice, corned beef, and white bread. The traditional Samoan diet was already heavy in saturated fats— more than 36 percent of the caloric intake was provided by coconut fats. The addition of canned meats, imported mutton, and chicken wings to the diet, accompanied by changes in lifestyle, has made obesity a significant problem among Samoans.

This problem is particularly acute in expatriate populations. The median weight of Samoan children in California falls close to the 95th percentile among the weights of other children there, which may have social as well as health implications for their assimilation within the broader American society. A large part of the weight gain among expatriate Samoan adults may be attributable to the weekend family feasts that are characteristic of Samoan culture. In the traditional culture, active individuals may have a caloric deficit most of the week, then balance it with the weekend *to'ona'i* feast. However, sedentary workers and other Samoans who participate economically in Western culture but retain the tradition of weekend feasting may be courting obesity, high blood pressure, and diabetes. Dietary surveys conducted by Joel Hanna and his co-workers at the University of Hawaii revealed that a typical sedentary Samoan worker consumes 3300 calories a day on weekdays and more than 5600 calories a day on weekends.

Changing diets can also alter the range of pharmacologically active compounds indigenous peoples ingest, since many cultures do not make a clear distinction between food and medicine. Just as a Westerner can drink tea both as a pleasant beverage and to calm an unsettled stomach, indigenous peoples value some foods for their medicinal as well as nutritive qualities. In Samoa, for example, a drink called *vaisalo,* which is consumed as a delicacy on ceremonial occasions by both sexes, is given to women immediately after they have given birth to a child. A rich, porridge-like drink that consists of coconut milk and grated coconut (*Cocos nucifera* [Arecaceae]), tapioca (*Manihot esculenta* [Euphorbiaceae]), and grated *vi* apple (*Spondias dulcis* [Anacardiaceae]), *vaisalo* is believed to restore the new mother's strength and ensure the expulsion of the placenta. Is *vaisalo* therefore a food or a medicine? Similarly, the Batemi or Masai people do not think of the cholesterol-lowering barks they add to their meat stews as medicine, though Western researchers are intrigued by the pharmacological potential of the compounds the barks contain.

Of plants used both as food and as medicine, perhaps those that tend to protect against malaria have aroused the greatest interest among ethnobotanists. Unfortunately, the quinine derived from the *Cinchona* trees of South America is somewhat ineffective against some of the more virulent strains of malaria, which are also exhibiting resistance to Fansidar, Maloprim, and other antimalarials. But new research on antimalarial plants offers hope not only for explorers but, more important, for rural peoples in the tropics who cannot afford Western pharmaceuticals.

When the anthropologists Nina Etkin and Paul Ross of the University of Hawaii studied the plants consumed by the Hausa people of Nigeria, they found that the bulk of the Hausa diet was provided by sorghum (*Sorghum bicolor* [Poaceae]), millet (*Pennisetum americanum* [Poaceae]), cowpeas (*Vigna unguiculata* [Fabaceae]), and peanuts (*Arachis hypogaea* [Fabaceae]). The porridges made from these grains, however, are consumed with soups made from a variety of other plants. Etkin and Ross have collected 61 semiwild plants consumed by the Hausa, which, although they account for only 3 percent of the

Hausa's caloric intake, contain a plethora of pharmacologically active compounds. The Hausa gather these "medicinal foods" throughout their environment: 64 percent come from their farms, 10 percent from their farm borders, and 26 percent from public lands. Etkin and Ross have found some of these compounds to be active against malaria, and they hypothesize that the plants function as malaria prophylactics.

Thus as indigenous peoples become assimilated within Western societies, not only do their sources of nutrition change, but their concepts of food and medicine become irrevocably altered as well. The Batemi and Masai who live in Nairobi have reduced their use of medicinal barks in the preparation of meat, and the Samoans who live in California no longer serve *vaisalo* to new mothers. Ethnobotanists find it very interesting to study cultures that are undergoing a profound change in lifestyle. Often such transitions reveal in striking detail the consequences, both cultural and nutritional, of different ways of life. A culture in transition provides in some sense an experiment whose historical precedent constitutes its own control. Yet often contemporary cultural changes follow a familiar pattern: indigenous peoples lose their knowledge of plants and adopt Western ways. The vesting of goods and services with monetary value is particularly corrosive of traditional ways. Many traditional uses of plants, such as the making of dyes and textiles, cease to have cultural importance and are retained only if they can generate cash. Since traditional ways commonly disappear when indigenous cultures encounter the West, ethnobotanists sometimes need to reach into the distant past for a more complete record of cultural transitions. One of the more fascinating sagas in the relationship between plants and people is the development of maize as a crop. Maize had profound consequences for every culture that adopted it.

The Development of Maize

Maize *(Zea mays)* was developed in Mexico as early as 3000 B.C., although ancestral forms found in caves near Puebla date to 5000 B.C. The origin of maize is controversial, but the plant probably developed from teosinte, *Z. mays* subsp. *mexicana,* a wild grass that still grows in Mexico. Maize is monoecious—that is, it bears both male and female flowers. The male flowers are borne in a tassel near the top of the plant. The one-seeded female flowers are borne lower on the plant in clusters, which, after pollination, mature to form cobs. Each female flower produces a long style known as corn silk, which functions to receive wind-dispersed pollen. Maize yields starch, but maize starch cannot be used for leavened bread because of its low gluten content: it lacks the culinary glue to

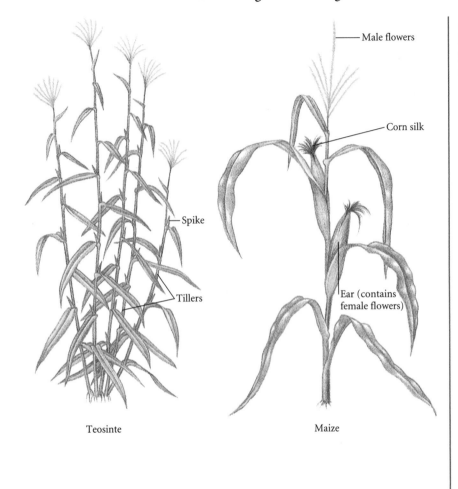

Teosinte

Maize

Zea mays subsp. *mexicana,* commonly known as teosinte, is believed to be the progenitor of domesticated maize (called corn in the United States), *Zea mays.*

hold a loaf together. The ground endosperm of the seeds, known as cornstarch, also can (after hydrolyzation) yield corn syrup, rich in glucose.

A brief stroll through any Central American market is sufficient to convince one of the tremendous diversity of maize types. Popcorn has water-filled cells in the endosperm (the starchy reserve of the seed) that explode when heated. Flour corn has a soft starch that makes a paste ideal for frying or baking into tortillas, or small flat cakes. Flint corn has hard starch and will not produce paste. Dent corn, which constitutes most of the current U.S. crop, has hard starch outside but soft starch inside. It is an ideal food for livestock. Sweet corn has sugar rather than starch in the endosperm and is served as "corn on the cob." But maize is remarkable not only for its variety but also for its sheer productivity.

Like all green plants, those that rely on C-4 photosynthesis take needed electrons from water molecules in the chloroplasts, releasing oxygen as a waste product. But in C-4 plants, unlike other plants, carbon atoms are transferred from the mesophyll to the airtight bundle sheath cells, where the enzyme rubisco is protected from oxygen poisoning. As a result, C-4 plants such as rice and maize are extraordinarily productive.

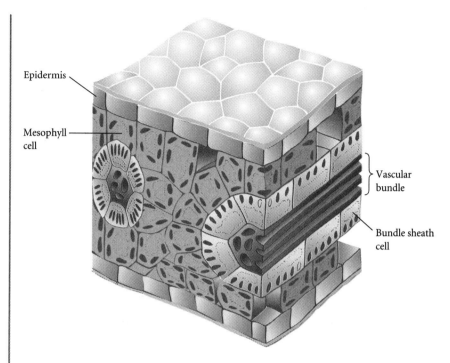

Epidermis

Mesophyll cell

Vascular bundle

Bundle sheath cell

Early Mesoamericans attributed the productivity of maize to supernatural sources, for in the right circumstances a primitive agriculturalist could produce far more food per hectare of maize than of any other crop. But the secret to maize's productivity lies not in supernatural forces but in its molecules.

A plant splits water into hydrogen and oxygen to obtain the electrons it needs for photosynthesis. Later it uses the hydrogen ions to form NADPH2, one of the molecules it needs for energy, but the oxygen produced in the process is spewed into the atmosphere. This is an extraordinarily fortunate thing for people and other animals, since the gaseous waste of plants is what we breathe. Plants, however, take in carbon dioxide. During photosynthesis the carbon atoms from carbon dioxide are used to build six-carbon sugars within the plant. An enzyme, rubisco (ribulose-1,5-bisphosphate carboxylase) catalyzes this reaction. Because of its importance in photosynthesis, rubisco is arguably the most important enzyme in the world. Unfortunately, though, oxygen competitively binds to the enzyme, stopping the reaction. Just as human beings are tormented by the by-products of industrial production, plants suffer from the presence of their primary waste product, oxygen, which interferes with carbohydrate formation at the cellular level.

Some plants, however, have evolved a photosynthetic mechanism that prevents oxygen from inadvertently poisoning rubisco. These plants separate the places where they split water atoms into hydrogen and oxygen from the place where they build sugars from carbon atoms. Water molecules are split, as in all plants, in the chloroplasts, but the rubisco is sequestered within airtight tissues in the center of the leaf.

The secret to the process is that in these plants, oxygen and all other atmospheric gases are excluded from the cells containing rubisco—the so-called bundle sheath cells surrounding the vascular bundles of the leaf. Carbon dioxide, rather than being imported as a gas, releases its carbon atoms to an intermediary four-carbon molecule, which crosses to the membranes of bundle sheath cells. Hence the entire process is called C-4 photosynthesis. Such C-4 plants as sugar cane, rice, and maize are among the world's most productive crops.

Every culture that has encountered maize has been radically changed by it. Indeed, much of the post-Columbian explosion of European populations was driven by the introduction of two New World crops: potatoes and maize. Before they changed Europe, they changed the cultures that first developed them. One of the most striking cases is the ancient Anasazi culture. After acquiring maize, the Anasazi abandoned the hunter-gatherer way of life for agriculture and became a highly complex, even urbanized society.

Did the Anasazi Domesticate Maize or Did Maize Domesticate the Anasazi?

The two-story Cliff House ruin in Mesa Verde, Colorado, is barely discernible from a distance. W. H. Jackson, a photographer for the U.S. Geological Survey, almost missed it in 1874 as he searched for a fabled Anasazi city. Built of stone approximately the same color as the red sandstone overhang that shelters it, Cliff House easily escapes detection by all but the most careful observer. Just at dusk, as Jackson stood heating coffee over his campfire, he looked up and saw several stone structures within a high, shallow cave, standing out in bold relief in the rays of the setting sun. Scrambling upward, Jackson discovered a ruin that once had sheltered five or six Anasazi families. Inside the stone structures were maize cobs, pottery, grinding stones, and cutting implements. It was a spectacular find, and Jackson spent most of the next day carefully photographing the ruins and their contents. Breaking camp on the same alluvial fan in Mancos Canyon that had once been cultivated by Anasazi families, Jackson re-

W. H. Jackson took this photograph of the
Cliff House ruin in 1874. Jackson missed
even more impressive ruins, and one of the
most significant archaeological finds of the
century, only just a few minutes up a side
canyon.

turned with his photographic treasures, confident that he had cracked the secret of Mesa Verde.

Had Jackson ventured only four more miles up a nearby side canyon, he
would have discovered something far more impressive than Cliff House: an entire walled city composed of three- and four-story buildings. Just one complex,
Cliff Palace, contains more than 200 houses and 23 kivas, or ceremonial cham-
bers. This and other nearby complexes—Balcony House, Spruce Tree House,
Square Tower House—were home to thousands of Anasazi Indians and represented the high point of their culture. W. H. Jackson missed making one of the
most staggering discoveries of North American archaeology by the narrowest of
margins. Discovery of this Anasazi metropolis would have to wait another 14
years before it was found by two cowboys of the Wetherill family as they chased
after lost cattle.

What happened to the residents of Mesa Verde? Why were its massive ruins
littered with intact baskets, pottery, blankets left as if the thousands of inhabitants had suddenly vanished in a single moment? The site had been continu-
ously occupied by the Anasazi Indians for more than a thousand years. The final multistory structures had been constructed about the same time that a
famous edifice was built in another urban agrarian society, the Tower of Lon-

don. And, just as the residents of London were decimated by the great plague of 1348, the residents of Mesa Verde had been the victims of a biological calamity only a few decades earlier. The consequences in both cases were grim, but the aftermaths were striking in their differences: nearly 50 percent of the inhabitants of London perished in the plague, but not a single person remained in Mesa Verde. The difference was that the bubonic plague killed only people, whereas the drought that descended on Mesa Verde in 1276 and lasted until 1299 not only killed people but destroyed their major crop.

In many respects the story of the rise and fall of the Anasazi culture mirrors these people's increasing reliance on maize. The arrival of maize at the Colorado Plateau transformed the Anasazi from hunter-gatherers into an urban agrarian society with a population density that exceeded those of major European cities at that time. The development of maize by the Anasazi chroni-

The Cliff Palace at Mesa Verde contained more than 200 family dwellings and 23 kivas. Before it was abandoned by the Anasazi in the late thirteenth century during a 23-year-long drought, Mesa Verde rivaled European cities of the time in population density.

cles both the immense possibilities and the real liabilities inherent in a shift from the hunter-gatherer way of life to agriculture. Since the Anasazi vanished from Mesa Verde in the late thirteenth century, we do not have their own account of their life and times. Still, by analyzing their ruins and rubbish piles and the climatological data provided by studies of tree rings, we can plot the trajectory of Anasazi culture in surprising detail. These analyses reveal how intimately maize was associated with the rise and fall of the Anasazi.

The development of the Anasazi culture can be divided into several distinct periods. During the Archaic period (5500–100 B.C.), the Anasazi were hunter-gatherers. They survived largely on the roasted seeds of Indian rice grass (*Oryzopsis* spp. [Poaceae]), cattails (*Typha latifolia* [Typhaceae]), saltbush (*Atriplex canescens* [Chenopodiaceae]), and sheep sorrel (*Rumex acetosella* [Polygonaceae]). Rabbits and occasional deer provided the bulk of protein in their diet. In the Archaic period the people lived in caves, under sandstone overhangs, or in depressions covered with timbers of juniper (*Juniperus scopulorum* or *J. osteosperma* [Cupressaceae]) and surrounding foliage, which provided roofing. Population densities of the Archaic Anasazi people were relatively low but stable. Organized warfare was uncommon.

Around 1000 B.C., however, with the arrival of the first maize plants from Mesoamerica, the Archaic way of life began to change. At first the Anasazi were slow to adopt maize as much more than a novelty. The ears were small and unproductive. The plants were very vulnerable to drought. Furthermore, maize cultivation required a sedentary lifestyle that was incompatible with hunter-gatherer ways.

As the centuries advanced, the Anasazi increased their cultivation of maize as they became more comfortable with agriculture. Around 100 B.C. they made the transition to what we now call the Basket Maker II period. Carefully woven, ornamented baskets characterize this period, together with sandals made of juniper bark and nets woven from yucca (*Yucca baccata* [Agavaceae]) fiber. Pit houses covered with timbers became more common. Protein was provided, as in the past, by hunting. Occasionally they downed even large game such as deer with sharpened flint projectile points mounted on slender shafts (*Phragmites australis* [Poaceae]) hurled from throwing sticks. The remains of squash (*Cucurbita* spp. [Cucurbitaceae]) and small-eared maize in Basket Maker II food caches suggest that agriculture was becoming a more important feature of Anasazi life. Although maize was probably a welcome addition to the wild-gathered grains, it could not constitute the sole means of sustenance for the Anasazi because it lacks lysine, an amino acid essential for humans. The advent of beans (*Phaseolus vulgaris* and *P. acutifolius*) in the Basket Maker III period (A.D. 400–700), however, made it possible for the Anasazi to attain a complete complement of amino acids solely from agriculture. The *Phaseolus* beans originated in Peru about 5500 B.C. and spread slowly through the Americas, following the route previously taken by maize. With the advent of the bean/maize combination, the Anasazi began to focus increasing efforts on agriculture, with particular emphasis on selecting maize cultivars with larger ears.

The increase in the quality and productivity of the Anasazi's agriculture led to dramatic changes in their culture. As they adopted a more sedentary lifestyle

As maize surpluses grew, and the Anasazi population soared, Anasazi material culture blossomed. Pottery became more complex, as potters used more sophisticated techniques and motifs. The dipper bowl from the Pueblo II period features the Anasazi mythical figure of music and virility, Kokopelli.

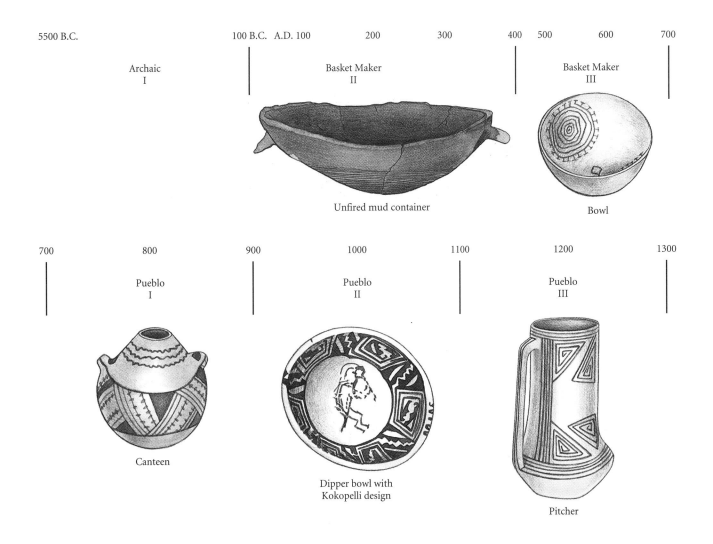

5500 B.C.

Archaic
I

100 B.C. A.D. 100   200   300   400   500   600   700

Basket Maker
II

Basket Maker
III

Unfired mud container

Bowl

700   800   900   1000   1100   1200   1300

Pueblo
I

Pueblo
II

Pueblo
III

Canteen

Dipper bowl with
Kokopelli design

Pitcher

in villages, they began to experiment with diverting water from streams and creeks to irrigate their maize plantings. Better agricultural techniques and careful selection of maize cultivars produced yields that not only fueled an increase in population density but in good years even generated surpluses. New hunting technologies, including bows and arrows, were introduced.

The combination of a more sedentary lifestyle, an increase in population density, and intermittent surpluses led to a flowering of Anasazi art forms at the beginning of the Pueblo I period (A.D. 700–900). Kokopelli, the Anasazi mythical figure of music, appears with increasing frequency in petroglyphs. The quality and variety of Anasazi basketry increased, and pottery developed into more ornamental forms with elaborate designs. Cotton was woven into cloth. Dwellings were built of stone above the ground. The pit houses of earlier periods were transformed into kivas to be used for ceremonial purposes. The maize surpluses, however, also led to increased conflict and warfare: hunter-gatherers rarely stored surpluses beyond their short-term needs, but hoards of maize made attractive targets for looters and pilferers.

Possibly in response to the increased need for protection from marauding bands of robbers, Anasazi families began to migrate to large settlements that resembled city-states during the Pueblo II period (A.D. 900–1100). By the beginning of the Pueblo III period (A.D. 1100–1300), Anasazi experiments in architecture had reached their apogee. Few contemporaneous European dwellings housed more people per square meter than did the multistory Anasazi complex at Mesa Verde, which served as home to thousands of people. With the multistory wood, mud, and stone dwellings initiated in the Pueblo III period, Mesa Verde experienced a flowering of material culture, art, and religion. The Anasazi may have fed maize to captive turkeys, generating not only meat but also feathers for blankets. Though the Anasazi's material culture reached its peak at Mesa Verde, urbanization brought epidemic diseases, the ever-present specter of famine, and raids by hostile tribes, and their life expectancy plummeted: only a fraction could expect to live beyond adolescence. Twenty-three years of relentless drought forced the Anasazi to abandon Mesa Verde and their other large settlements and to migrate south to better drainage areas. Today the descendants of these people include the Zuni, Hopi, and Rio Grande Pueblo tribes.

As the history of the Anasazi demonstrates, the development of agriculture can transform far more than a society's diet. Usually it increases the mean level of food production but reduces resistance to unpredictable environmental impacts such as drought. Once surpluses accumulate, not everyone's labor is needed in the fields or to procure food, and a socially stratified society becomes likely.

When a society produces agricultural surpluses, the surplus can support artisans, warriors, and ruling classes whose primary pursuits are not agricultural. Laborers can devote their primary efforts to public works rather than to foraging. A large structure such as the Sphinx in Egypt, the 216-foot Pyramid of the Sun at Teotihuacán, Mexico, or the temple complex at Angkor Wat in Cambodia is simply beyond the means of any hunter-gatherer society to produce: it could be constructed only by a society that produced surpluses of a productive crop such as wheat, maize, or rice.

Thus while the development of agriculture is likely to subject a society to famine, warfare, and the problems associated with urbanization, the intermittent surpluses it generates can also lead to a flowering of the material culture and significant social stratification. Patrick Kirch, an ethnobotanist at the University of California at Berkeley, has suggested that Polynesian agriculture went through a sequence of three discrete stages—colonization, development, and intensification—and that this progression led inexorably to the rise of chiefly classes there. But chieftains could effectively translate agricultural surpluses into political power only if those surpluses were not perishable. People can maintain livestock surpluses through the simple expedient of keeping animals confined but alive until they are ready to slaughter them. For surplus crops to become the basis of political power, however, they must be kept from spoiling. In an arid climate, grain can be stored in vermin-proof pottery for a long period; but in the humid tropics, protection of food from mold and rot requires far more ingenious solutions.

Polynesian Strategies for Food Preservation

In 1805 the Russian navigator A. J. von Krusenstern visited the island of Nukuhiva in the Marquesas. Krusenstern found the people eating a "sour pudding" as a staple. "From ten to fifteen paces from their houses are several holes, paved with stones and covered over with branches of trees and leaves," Krusenstern wrote.

> In these they keep their provisions, consisting chiefly of baked fish and of sour pudding, a kind of dough made of the taro root and breadfruit. . . . Their chief dish is this sour pudding, which is not disagreeable and may be compared to an apple tart. . . . Their manner of eating is highly disgusting; they snatch up the sour pudding with their fingers, and carry it with great greediness to their mouths.

Doughlike breadfruit, *Artocarpus altilis,* became a major staple of the Polynesian people. Although the fruit is plentiful when in
season, Polynesians developed pit fermentation techniques to extend the period of
its availability.

Although Krusenstern obviously disapproved of Marquesan table manners,
he was intrigued by this "sour pudding" that the Marquesans fermented in pits.
Some pits "paved with stones" were very large. Krusenstern observed one in the
Atu Ona valley that was 25 feet deep. A later anthropologist in the Marquesas,
Ralph Linton, described a "sour pudding" pit that was 30 feet deep and 18 feet
in diameter.

The Marquesan islanders called the sour pudding *ma* and accorded it such
great importance that they cast the first crop of breadfruit into the *ma* pits. The
first crop of breadfruit, Linton wrote,

> belonged entirely to the chief, and served to fill his private breadfruit
> paste *(ma)* pits from which his household, guests, assistants, and
> workers were fed, and the great tribal reserve pits in the back of the
> valley, which were filled in good times against famine. The second
> harvests were used to fill the private family pits. . . .
>
> The horror of famine must have loomed large in the minds of the
> Marquesans, for, in spite of their happy enjoyment of the present
> time, without a thought for the future in other respects, they devel-
> oped in the making of *ma* a very elaborate system of conservation of
> breadfruit in silos in the ground.

Although *ma* is tart, many European visitors learned to enjoy it. Indeed, no less a literary luminary than Herman Melville ate *ma* for an entire month with the Marquesans after he deserted his ship in Nukuhiva in 1841. He recorded the experience five years later in his first book, *Typee:*

> This staple article of food among the Marquese islanders is manufactured from the produce of the breadfruit tree. It somewhat resembles in its plastic nature our bookbinders' paste, is of a yellow color, and somewhat tart to the taste. . . . This kind of food is by no means disagreeable to the palate of a European, though at first the mode of eating it may be. For my own part, after the lapse of a few days I became accustomed to its singular flavor, and grew remarkably fond of it.

Ma functioned as far more than a delicacy in Marquesan society. First, *ma* allowed the people effectively to extend the brief fruiting period of breadfruit, when the fruit is so abundant that much of it rots on the ground before it can be consumed. Second, and more important, the fermented breadfruit paste allowed the Marquesans to survive a prolonged famine. Breadfruit (*Artocarpus altilis* [Moraceae]), taro (*Colocasia esculenta* [Araceae]), yams (*Dioscorea alata* [Dioscoreaceae]), sweet potatoes *(Ipomoea batatas),* and other Polynesian crops, unlike most grain crops, are not annual plants propagated by seed but perennials propagated by cuttings. Thus destruction of the standing crop by drought, volcanic dust, cyclone, or warfare could destroy any hope of a future crop. Even if some cuttings could be salvaged, it takes up to seven years for a breadfruit tree to produce fruit. Perhaps it was for this reason that in time of war, the chief targets of the Polynesians' attacks were their enemies' crops.

Accounts of crop destruction as a major objective of military sieges are common in the journals of early European visitors to Polynesia. "Observing the mountains surrounding the valley to be covered with numerous groups of natives, I inquired the cause," wrote Captain John Porter, who visited Nukuhiva 18 years after Krusenstern.

> [I] was informed that a warlike tribe residing beyond the mountains had been for several weeks at war with the natives of the valley, into which they had made several incursions, destroyed many houses and plantations, and killed a number of breadfruit trees by girdling. . . . In the afternoon several officers went on shore to visit the villages, when I perceived a large body of Happahs descending from the mountains into the valley among the breadfruit trees, which they soon began to destroy.

The anthropologist Ralph Linton photographed these *ma* pits used for fermenting breadfruit in Hivaoa, Marquesas Islands, in 1920.

The *ma* pits allowed fermented breadfruit to be maintained for months and even years, thus bridging the gap between the destruction of one crop and the production of another. It should come as no surprise, then, that the islanders often built large communal stone *ma* pits high on ridges or inside of fortifications, so that they were prepared to withstand a siege.

The technology for producing fermented breadfruit paste was not confined to the Marquesas. The paste was called *mahi* in Tahiti, *ma'i* in Mangareva, *masi* in Samoa, *me* in Tonga, *maratan* in Ponape, *manakjen* in the Marshall Islands, and *namandi* in Vanuatu. Enjoyed in times of plenty and depended on in times of scarcity, fermented breadfruit, taro, or banana paste allowed agricultural products to be preserved for years.

The process of making a fermented breadfruit pit is exacting in its requirements. Chief Ofala Va'alaufuti of the village of Falealupo on Savaii Island, Western Samoa, explained that the most important consideration during construction of a *masi* pit is to make sure that it is *lē tolofia* (airtight); otherwise "everything inside will rot and be badly spoiled." Va'alaufuti lines the pit with about 50 large, waxy leaves of *Heliconia laufao* [Heliconiaceae], the blades overlapping, so that when the leaves are folded over the breadfruit, they form a large airtight pocket. The breadfruit are washed and scraped, placed in the pit, covered with leaves, and buried with dirt and rocks.

Under the direction of Ofala Va'alaufuti an experimental *masi* pit in Samoa was constructed in which the breadfruit was allowed to ferment for 34 days. Since years had passed since a *masi* pit had been constructed on the island of Upolu, some villagers were concerned that the breadfruit would rot, but when the pit was unearthed, only a sweet-smelling paste was found in it. Unlike *ma,* which the Marquesans eat as a paste, Samoan *masi* is combined with coconut cream and baked to produce a breadlike loaf. Baked *masi* has a strong fermented taste something like sauerkraut or Limburger cheese. The month that the breadfruit fermented in the trial pit was a very short time; some villages have unearthed *masi* pits after 20 years to find their contents still edible.

Although the arrival of Western food has caused breadfruit fermentation to vanish from many parts of Polynesia, Western technology has been incorporated into the process in the Manu'a Islands: many chiefs now line their *masi* pits with plastic tarpaulins, and others conduct the entire fermentation process inside plastic kegs that originally contained corned beef.

Famine Foods and Sago Production

The fermentation of starchy crops in pits can effectively stretch the length of time food is available, but what happens during a famine so prolonged that even the *masi* pits are inadequate? Interviews with villagers in Falealupo who survived a famine after a series of storms destroyed their crops reveal that they have a set of famine foods that are completely distinct from crop plants.

None of the plants in the table on page 86 are indigenous to Samoa: all arrived with the original Polynesian colonizers, and nearly all of them persist today in the wild. Only an agrarian society would have to rely on introduced rather than indigenous plants in times of famine: hunter-gatherers know how to use the numerous plants indigenous to the area. The table confirms that the first Polynesians to settle in Samoa brought agriculture with them and did not develop it after they arrived there.

Why do the Polynesians not eat these plants as part of their normal diet? An examination of the plants reveals that most of them were prototype cultivars discarded in favor of new and improved models. The presence of nut crops such as *Inocarpus fagifer* [Fabaceae] and *Terminalia catappa* [Combretaceae] may hark back to the long-forgotten days of the cultivation of *Canarium* [Burseraceae], a nut crop introduced by the Polynesians' ancestors. Similarly, denigration of the seeded banana *Musa acuminata* [Musaceae] by the appellation *tae manu* (literally "animal feces") disguises its former importance in the evolution of more recent seedless banana cultivars. Primitive seeded forms of Polynesian crops such as bananas and breadfruit can persist in a feral state, but the more "advanced" seedless forms require active human intervention to be perpetuated. As any nursery worker knows, it is far quicker to raise plants from

Fruits of the wild seeded banana *Musa acuminata* subspecies *banksii*, found in Samoa. Denigrated by the local people, seeded bananas are eaten only during periods of extreme famine.

Wild Plants as Samoan Famine Foods

Species	Family	Samoan Name	Edible Part	Preparation
Monocotyledons				
Cordyline terminalis	Agavaceae	Ti vao	Rhizomes	Cooked
Cyrtosperma chamissonis	Araceae	Pula'a	Rhizomes	Cooked
Dioscorea alata	Dioscoreaceae	Ufi vao	Roots	Cooked
Dioscorea esculenta	Dioscoreaceae	Ufi lei	Roots	Cooked
Musa acuminata	Musaceae	Tae manu	Fruits	Cooked
Metroxylon warburgii	Arecaceae	Niu lotuma	Stem	Starch extraction
Cocos nucifera	Arecaceae	Niu	Meristem	Cooked/raw
Tacca leontopetaloides	Taccaceae	Māsoā	Roots	Starch extraction
Dicotyledons				
Terminalia catappa	Combretaceae	Talie	Seeds	Cooked
Inocarpus fagifer	Fabaceae	Ifi	Seeds	Cooked
Adenanthera pavonina	Mimosaceae	Lōpā	Seeds[a]	Raw
Syzygium samarangense	Myrtaceae	Nonu fi'afi'a	Fruits	Raw

[a] Early European introduction.

cuttings than from seed, and as any child knows, seedless fruit is easier to eat. But selection for efficiency resulted in increased vulnerability to famine because seeds are more easily stored than cuttings.

Thus the old, discarded crop plants that persist in the wild are denigrated until a famine strikes; once the *masi* pits are emptied, what has been derided is then treasured and eaten.

In this light, the presence of the sago palm, *Metroxylon warburgii* [Arecaceae], among Samoan famine foods is most interesting. The Samoan name for the palm, *niu lotuma* (Rotuma palm), seems puzzling, for Rotuma is a small isolated island 760 miles to the west of Samoa. Even more puzzling is the use of the palm as a famine food. Sago starch is a staple in Papua New Guinea, Irian Jaya, and parts of Indo-Malaysia, where the palm is called *sadu, sagu,* or some cognate. Although the genus was anciently introduced to Samoa, the ancestral name did not persist, perhaps because people ceased to rely on it for food unless famine struck.

In an evolutionary sense, sago palms are the botanical equivalents of salmon. With a "monocarpic" life cycle, sago palms live almost their entire lives without reproducing, flowering only immediately before they die. That single explosion of flowers borne on an erect cluster produces upwards of a million blossoms. Just as a city on a hill cannot be hid, so the massive erect inflorescences of *Metroxylon* can scarcely evade notice, particularly by the swarms of insects that pollinate them. Indeed, it is possible to interpret the entire 6-to-15-year life span of a *Metroxylon* palm as preparation for that single burst of flowering, which attracts pollinators from kilometers away. Each year of photosynthesis the palm sequesters more starch within its trunk. At the end of the palm's life,

Sago starch from the trunk of *Metroxylon* palms is a staple throughout the shaded area on the map. In Indo-Malaysia and New Guinea, its harvest is accompanied by complex rituals. Moving eastward through the Pacific, the rituals become simpler until we reach Rotuma Island, where no ceremonies are performed with the palm's harvest. Palms of the genus grow in Polynesia east of Rotuma, but the Polynesians eat sago starch only in times of famine.

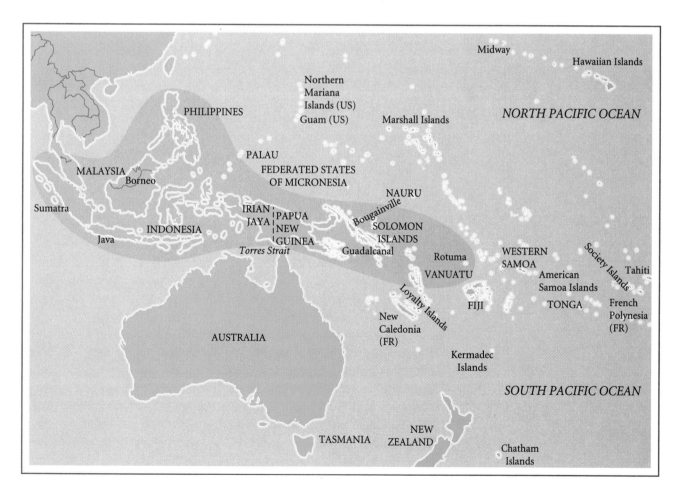

A sago palm is being harvested by Rotuman Michael Stevens, who will afterward extract the starch from the trunk.

that precious store, accumulated day by day over many years, will be called upon to fuel a single burst of flowering.

Indigenous peoples have become expert at predicting when the starch stored in a sago palm is at its maximum and have devised technologies to extract it. Although groups in Indo-Malaysia extract sago starch daily, the Polynesians eat it only in times of famine. Thus it is interesting to discover that the people of Rotuma, the small equatorial Polynesian island for which the Samoan plant is named, still harvest sago starch on a regular basis.

Ethnobotanist Will McClatchey found that the sago palm, known as *ota* in the Rotuman language, has a variety of uses. The leaves *(rau ota)* are used for house thatch, and the rachis (main axis) of the leaves provides material for making brooms. The immature fruits *(hue ne ota)* are eaten raw after removal of the pericarp, the outer skin of the fruit. And, of particular interest, the trunks are used to prepare starch.

The preparation of *Metroxylon* starch, called *mar ota*, involves several stages. First, the tree is chosen, preferably one that is just about to flower, so that the maximum amount of starch will be available. After the tree is felled, the trunk is split longitudinally with a single well-placed blow of an ax. A half of a coconut shell that has been sharpened along the edge is used to scrape out the tissue from the inner part of the trunk. The grated pith from the trunk is then placed in a clean fabric bag. Before cloth was introduced by Europeans, the Rotumans used the textile-like bases of coconut fronds or fibers from banana petioles (the stalks of the leaves). The bag is placed in water in a container. After the gratings are wrung out, the cloudy water is allowed to settle. When the water is then poured off, the remaining white, sludgelike layer of starch is dried.

Metroxylon starch is used in a variety of ways. It serves as a thickening agent in soups and stews and as starch for clothing. It is also the main ingredient of some foods. In one of its more popular forms it is baked or fried with coconut milk and sugar to produce a confection known as *fekei mara,* which has a sweet, delicate taste resembling that of tapioca.

The discovery that the Rotumans make such extensive use of sago starch raises a question: Why do the rest of the Polynesians not use sago, if their ancestors originated in or passed through areas of the world that depend heavily on it? There are several possibilities. (1) The Indo-Malaysians and Melanesians may have discovered the uses of sago after the Polynesians migrated. Or (2) the knowledge may have been lost or the palm replaced by another starch source that is easier to prepare or quicker to grow. Or (3) the Polynesians' ancestors may not have taken the *Metroxylon* species with them, or it may not have fitted the ecological template of the islands when they arrived.

The first suggestion is unlikely to be correct, since sago was probably used in Indo-Malaysia and Melanesia before the migration to Polynesia and may even have been one of the foods used by the earliest inhabitants. The second possibility, loss of the knowledge required to harvest the crop, would require replacement by a superior source of starch. Possible alternates are Polynesian arrowroot (*Tacca leontopetaloides* [Taccaceae]) and sweet potatoes. Both are easier to process and—more important from a colonist's point of view—quicker to grow than *Metroxylon* palm. Furthermore, development of the edible aroids (including taro) and breadfruit and the invention of pit fermentation techniques may have reduced the colonizers' reliance on sago starch. As far as the third possibility is concerned, the Samoans' and Tongans' claim that *Metroxylon warburgii* was introduced from Rotuma corroborates the view that *Metroxylon* was not part of the cultivar inventory carried by the early colonists of Samoa and Tonga. A fourth alternative focuses on the fact that Rotuma is the easternmost point at which a species of *Metroxylon* is used as a source of starch. Perhaps the intervention of the Europeans interrupted a movement of the sago palm and its use as a food source from Melanesia into Polynesia.

Which of these alternatives is correct is at present unknown. In any case, sago production in Polynesia represents a very interesting case of human dispersal of a crop eastward among the Pacific islands, with gradual attenuation of knowledge concerning its use and importance. In Indo-Malaysia the harvesting of sago is associated with complex rituals, none of which are practiced in Polynesia. The processing techniques, but not the rituals, are maintained in Rotuma, but the Samoans only vaguely know that the palm is edible and have forgotten how to process it efficiently.

Crop Origin: From Myth to Biotechnology

The origin of new crops is, in most cultures, shrouded in myth and legend. Samoan legends assert that the plant most important to Polynesians originated from a dying freshwater eel that had pursued the maiden Sina from island to island. "I love you and wish to provide you a final bequest," the eel told Sina. "Bury my body, and a plant will grow. I will caress you each time you drink its fruit." Sina was faithful to the wishes of the dying eel. The young shoot that emerged looked very eellike. It grew into a tall and unbranched tree of unparalleled utility to the Polynesian people: the roots are used for medicine; the leaves are used to weave thatch and baskets; the trunk provides timber for construction; the outer husk of the fruit is used for firewood and for cordage fiber; the shells of the seeds are used to make cups and other containers; the meat of the seeds can be eaten raw, pounded into a thick cream for cooking, or baked in the sun to produce a rich oil for ointment or illumination. And one sees the two eyes and mouth of an eel each time one drinks the sweet and sterile liquid of a coconut.

By the word "myth" we do not mean that the explanation is false, but rather that it expresses the way people perceive the universe and their place in it. This is not to suggest that all indigenous legends are devoid of useful scientific content: even though Western scientists may find it difficult to believe that a Navajo deity created maize in what is now the southwestern United States, independent of any Mesoamerican origins, Samoans' claims about the introduction of the sago palm to their islands from Rotuma have in fact been verified. Some of the same crops that inspired the myths of indigenous societies have now become staples of Western society; ethnobotanists are intrigued by human experience with crop exchange, and the major changes it produced, in all human societies.

Whereas the sources of indigenous crops may be cloaked in myth, the biological exchange that resulted from Columbus' voyages to the New World can be precisely dated; yet its effects are still being calculated. The introduction of New World crops to the Old World changed not only people's diets but the cultures of entire nations. Who can imagine Italian cuisine without tomatoes, Scandinavian cooking without potatoes, or West African feasts without manioc? Alfred Crosby of the University of Texas has argued that the introduction of maize and potatoes alone led to a doubling of Europe's population in post-Columbian times. But this crop-induced population explosion ultimately had negative demographic consequences as well. One has only to think of the Irish potato famine of 1845–46 and the subsequent Irish migration to North America to realize that the fate of nations resided in the precious seeds and roots that Columbus carried back with him to the Old World.

As we mentioned earlier, maize lacks the essential amino acid lysine. A diet based solely on maize will result in pellagra and other diseases caused by nutritional deficiency. This is why the Anasazi needed both maize and beans to fuel their thousand-year culture. The lack of lysine in maize is also perhaps the reason that potatoes from the Andes were so appealing to the poor of Europe: eaten in sufficient quantity, potatoes can provide a complete complement of amino acids, including lysine. Van Gogh's painting *The Potato Eaters* is poignant because it depicts the Dutch underclass, but not because it portrays people in danger of starvation. In the depleted soils farmed by peasantry in northern Europe, potatoes were a godsend; they can also be grown in poor soil in areas where the growing season is short or left in the ground, if necessary, so they are less sensitive to the timing of the harvest than competing poor-soil crops, such as rye.

Maize, sweet potatoes, and other New World crops also fueled much of the population explosion of China, for more than a third of China's current food supply is provided by crops of New World origin. Crops of American origin— not only the staples of maize, potatoes, and manioc but also squashes, pumpkins, chile peppers, cacao, cranberries, and peanuts—provide nourishment and dietary variety to millions of people throughout Africa, Asia, and Europe, as well as the Americas. And now biotechnology promises to improve even that cornucopia of useful plants.

Plant breeders have produced mutant maize varieties that produce lysine and elevated concentrations of other amino acids. Efforts are also being made to produce potato seed that can be sown directly instead of the tubers that are now required. However, production of new varieties can also produce new liabilities; single-genotype maize that was commercially sown throughout the southeastern United States proved vulnerable to fungal attack in the late 1950s. To find disease-resistant genes for breeding better varieties, plant breeders must return to diverse ancestral stocks. But many of these plants are disappearing in all but the most remote regions. Why should a mountain farmer in Mexico continue to plant the low-yielding maize variety his grandfather grew when store-bought hybrid seeds can double his yield? And yet his fields, and others like them, may contain the only hope for the preservation of diverse strains that exist outside the supercold cryogenic freezers of several seed repositories in the United States, United Kingdom, and Russia. Indigenous peoples provide an incalculably valuable service to the rest of the world by keeping some of these rare genetic strains alive. But to fill this very modern need, plant breeders often have to turn to the very distant past.

New Crops from Old: The Case for Amaranth and Hemp

The annual tribute delivered to Aztec emperor Montezuma was massive: 10,000 baskets of a grain called *huauhtli* were carried from the outlying provinces to the pyramids near what is now Mexico City. The royal court consumed part of the offering but most of the harvest of the tiny seeds (each 1 millimeter in diameter) was given to the priests. In an elaborate ceremony they fashioned an image of the god Huitzilopochtli from red paste made from the ground grain and paraded the idol to the base of the great pyramid. There they broke up the image and distributed the pieces to the crowds, who believed it to be the body

Amaranthus, shown here being gathered from a field in Nepal, is now cultivated as a nutritious and important grain crop in many areas of the world.

and blood of their god. Disturbed by the obvious similarities of the ceremony to the Holy Eucharist, the early Christian fathers suppressed use of the grain: ceremonies involving *Amaranthus hypochondriacus* [Amaranthaceae] were permanently banned. Unfortunately for the conquistadors, biology worked against this European concept of reform, for *Amaranthus* has a highly efficient C-4 photosynthetic system, is extraordinarily hardy, and is easily cultivated with hand tools. It is similarly easy to harvest: the large inflorescences of the one-to-two-meter-tall plants are simply rubbed by hand to break off the tiny fruits, and the chaff is winnowed away in a basket. And *Amaranthus* provided one essential contribution to the maize-dependent Mexican societies: lysine.

That essential amino acid, so notably missing from maize and yet produced in such abundance by *Amaranthus,* ensured that this crop would follow the ancient trade routes from Mesoamerica, tracking the previous spread of maize to the north. Jonathan Sauer of the University of California at Los Angeles reports that *A. hypochondriacus* appeared in the Aztecan sequence around A.D. 500 and then spread as far north as the Colorado Plateau, where the Paiutes presented the explorer John Wesley Powell with both wild and cultivated *Amaranthus* seeds. Ancient hoards of *Amaranthus* have been found as far east as the Ozarks. Today the grain is still grown in remote areas of the Sierra Madre of Mexico, particularly by the Tarahumara Indians, who grind the seeds into a dough to make tortillas and tamales. In post-Columbian times two other species,

Amaranthus cruentus and *A. caudatus,* were introduced to the Old World and spread as far as India and Africa. Because of its protein content, *Amaranthus* and maize make a nearly ideal diet.

Though they eventually were displaced from Mexican fields by higher-yielding grains, *Amaranthus* species are now the subject of intensive scientific research as a crop for the developing world because of their hardiness, protein content, and ease of cultivation with hand tools. Scientists at the University of Nebraska have also conducted trials in attempts to develop an *Amaranthus* cultivar suitable for industrial cultivation. Because of its high protein content and significant productivity, *Amaranthus* may emerge as one of the most important new crops of the twenty-first century.

Food crops are not the only subjects of current searches. In a world of vanishing forest resources and increasing demands for fiber and wood pulp, significant attention is now being devoted to indigenous sources of fiber. Hemp (*Cannabis sativa* [Cannabaceae]), long a traditional source of fiber for rope, fabric, and even paper, was once extensively planted throughout the United States with government support. *Cannabis* is a species of many uses, with a history of applications for nonnarcotic purposes; its fibers were woven into fabric at least as early as 8000 B.C. The fiber, known as hemp, is so strong and durable that ship's sails were woven from it from the fifth century B.C. until the mid–nineteenth century. *Cannabis* was the major paper fiber until 1883, used to produce Bibles from Gutenberg to the King James version. Thomas Paine's pamphlets were printed on paper made from hemp, and the first and second drafts of the Declaration of Independence were written on hemp paper imported from Holland. In addition, cordage, lighting oil, building materials, and even plastic pipe have all been manufactured from the hemp plant.

THC, a compound found in hemp, is extremely valuable in the treatment of intraocular pressure from glaucoma as well as nausea and other unpleasant effects of cancer chemotherapy. The hemp crop, however, has been one of many casualties of the epidemic of illegal drug use in the United States. Planting hemp for any purpose within the boundaries of the United States is now illegal.

China, however, maintains a thriving hemp industry, which was recently studied by Robert C. Clarke of the International Hemp Association in Amsterdam. Hemp was one of the earliest crop plants in China; evidence of its cultivation there spans 5000 years. Bast fibers of the male plants were used to weave the cloth of which most Chinese garments were made in ancient times. The yarn woven from such hemp fibers was exceedingly fine, the equivalent of modern 70–80-count thread. Perhaps it was because of the fine texture of hemp thread that people of the Western Han dynasty (260 B.C.–A.D. 24) were buried

Hemp being harvested in China. The Chinese harvest more than 100 tons of hemp fiber annually in Shandong Province, to produce fiber for paper, rope, and cloth.

in hemp garments. During the Tang dynasty (A.D. 618–907) shoes were made from hemp. Of particular interest now is the ancient use of hemp fibers to produce paper. Indeed, what is claimed to be the oldest surviving piece of paper, found in a tomb near Xi'an in Shaanxi province, has been dated to 180 B.C. The potential to make paper from such a fast-growing, renewable source of fiber without cutting a single tree has seized the imagination of conservationists throughout the world.

Several varieties, including *lai wu* and *fei cheng,* are grown in Shandong province, where Clarke studied the traditional hemp industry. Seed is broadcast by hand, and then the fields are fertilized. At harvest the stalks are cut with a sickle at ground level and placed in water for "retting," a two-to-three-day process that clears the fibers of extraneous plant material. After retting, the stalks are stripped of fiber by hand and tied. Locally the fibers are braided together to make twine, rope, sacking, and burial cloths as well as crude paper. Most of the hemp, however, is produced for export. Clarke estimates that more than 100 tons of hemp fiber are produced annually in Shandong province. Some hemp fiber is shipped directly to Japan for specialty paper production, while other parcels of fiber make their way to paper and shoe factories in China.

There is interest in reintroducing hemp as a fiber for clothing, because of its many useful properties and the growing popularity of clothing made from natural fibers. The manufacture of hemp fabric is now a cottage industry with an annual value of $50 million—a far cry from the days when Thomas Jefferson grew it. Consumers may not always be aware when they are purchasing hemp products; the *New York Times* reports that hats made from hemp are even sold in Disney World's Indiana Jones gift shop. Yet some designers have conspicuously featured hemp in their products. Since 1984, Ralph Lauren has used it in some of his clothing lines, and Calvin Klein combined hemp fabric with linen in a 50–50 blend for his 1995 collection of duvet covers, decorative pillows, and pillow shams. Thus an ancient fiber crop, made illegal in many countries because of its abuse as a recreational drug, is now enjoying a resurgence of interest, particularly among chic designers. But hemp is not the only crop to be rediscovered; the palette of modern diners is being increasingly diversified by the reintroduction of very old crops.

Ethnobotany and Haute Cuisine

Upscale restaurants from Monterey to Manhattan serve the new American cuisine: blue-flour corn cakes, fried cactus pads, and perhaps a garnish of star fruit (*Averrhoa carambola* [Oxalidaceae]) or pine nuts (*Pinus edulis* [Pinaceae]). In the produce sections of most large American supermarkets kiwis (*Actinidia chinensis* [Actinidiaceae]), taro, and tamarind pods (*Tamarindus indica* [Caesalpiniaceae]) are now common features. Yet behind what appears to be merely a pursuit of culinary novelty is the result of a renaissance in ethnobotanical research on indigenous crops. Organizations as diverse as Native Seed Search in Arizona, The New York Botanical Garden, The Royal Botanic Gardens at Kew, the U.S. Department of Agriculture, and the Food and Agriculture Organization of the United Nations are urgently seeking new strategies to preserve the world's legacy of genetic and species diversity that constitutes the heritage of indigenous crops.

The attempt to preserve species diversity is a race against time. Each day crop varieties bearing precious genes give way to new "improved" hybrid strains distributed to rural farmers by well-intentioned agricultural improvement programs. International programs have been designed to provide "genetic arks" to try to preserve a remnant of this diversity against the modern flood of productive but genetically uniform crop varieties. The Kew Seed Bank, for example, at Wakehurst Place in Sussex, England, stores hundreds of millions of seeds at $-20°C$. Not all seeds, however, can survive freezing. Seeds of temperate zone trees such as oaks, and, more ominously, many rain forest trees die when they are frozen. Research is continuing at several centers around the world on means of preserving seeds, but precious little research has been done on techniques for preserving tropical crops that are cultivated not by seeds but by cuttings or tubers. Living germplasm collections, such as the sweet potato collection planted by the ethnobotanist Douglas Yen at Lincoln, New Zealand, and the breadfruit collection assembled by the ethnobotanist Diane Ragone at the National Tropical Botanical Garden in Hawaii, are vulnerable to changes in funding and administrative policies.

Still more insidious, however, is the disappearance of indigenous knowledge of crop varieties that have been lovingly developed over millennia. This loss is not the result of reasoned opposition to "primitive ways," but of their inadver-

Hundreds of millions of seeds, from thousands of species (including many with traditional uses), are stored in the Kew Seed Bank at Wakehurst Place in England. Kept in airtight aluminum and glass containers, at $-20°C$, the seeds will remain viable for decades and in some cases even centuries. The seed bank stores material from over 100 countries, making it one of the most diverse collections in the world. Another goal of the scientists at this facility is to better understand the physiology of seeds.

tent erosion. The gravest threat to indigenous diets is not malice, but often something as innocuous as Western fast foods.

Recently Paul Cox was teaching a short course in ethnobotany to 12 Minangkabauan students at Andalas University in Sumatra, Indonesia. One day the class stopped at a roadside café for a modest lunch. Traditional Minangkabauan dishes were quickly placed on the table. As a spur-of-the-moment exercise, Cox asked the students to count the number of plant species represented in their lunch. The tally soon emerged: 54 plant species, including six species gathered in the wild. As Cox explained to the students, a typical fast-food lunch in the United States of a hamburger and french fries represents only eight to ten plant species in anything more than trace amounts. None of the species are gathered wild. Despite the emergence of American haute cuisine, clearly the world culinary palette is becoming less colorful, and our increasing dependence on a narrower and narrower genetic base presages increasing vulnerability to crop pandemics. The message of the Anasazi is clear: there are efficiencies to be gained from depending on a basketful of only a few crops, but that efficiency sometimes comes at a terrible price.

The Tahitian canoes and sailing craft painted in 1773 by William Hodges of Cook's expedition had extraordinary maneuverability and speed, surpassing even that of the European sailing ships. But the Polynesians constructed far larger seacraft for long-distance voyaging and migrations. This use of plants could truly be said to have altered the course of world history.

Plants as the Basis for Material Culture

CHAPTER 4

Throughout the world plants are the basis of human material culture. We use plants to meet our most basic needs for food, clothing, and shelter. Most indigenous societies, which traditionally have lacked the metals and synthetic materials ubiquitous in Western society, rely almost entirely on plants for their material needs. The number and variety of uses to which people put indigenous plants is astonishing, ranging from woven cords and plant adhesives of sufficient strength to hold large oceangoing rafts together to arrow poisons that as hunting tools rival shotguns in their lethal efficiency. Some uses of plants, such as those used for bodily adornment, reflect indigenous taste more than necessity. But perhaps nowhere is the influence of plants on culture more dramatic than in their use to manufacture sea craft that transport people and their crops across vast stretches of the ocean.

The consequences of long-distance voyaging on cultures are clear: the North American continent is currently occupied and politically dominated by peoples who trace their ancestry to another hemisphere. European culture and economic systems have been replicated in outposts as distant as Australia and Africa. The biological consequences of long-distance voyaging are equally profound. The exchange of crops between the Old and New Worlds in the aftermath of the expedition of Columbus—namely, the inadvertent introduction of exotic weeds, parasites, and diseases—changed not only the course of human civilization but also forever altered the trajectory of biological evolution. We now know that significant plant exchanges occurred before Columbus. Phoenician seafarers sailed from the Middle East around the southern tip of Africa many centuries before the Christian era. Arabian traders voyaged along the eastern coast of Africa into the Indian Ocean. Balsa-wood rafts transported people along the western coast of South America and perhaps, as Thor Heyerdahl has argued, into the islands of the South Pacific. Exchanges of plant materials between the Old and New Worlds long before Columbus, once dismissed as fanciful, are now considered quite likely.

The evidence for prehistoric voyages of great distances poses an interesting question: How did indigenous peoples construct oceangoing vessels? Early European explorers drew sketches of primitive vessels they encountered in their journeys, yet only recently have ethnobotanists paid close attention to the plants used in the construction of boats and rafts. Studies with the few remaining indigenous shipwrights demonstrate that these craftsmen function much like modern aeronautical engineers: after considering the material constraints imposed by performance criteria for different parts of a craft, they then carefully select plant materials to meet those needs. The large oceangoing craft of early seafarers required special skills and special materials and so called for painstaking searches for appropriate plants.

Plants and Indigenous Voyaging

Throughout the world, people have used plants to construct vessels for navigating rivers, lakes, and seas. Some prehistoric ocean voyagers traversed great distances: in the tenth century Erik the Red succeeded in voyaging more than 800 miles (1287 km.) from Iceland to discover Greenland. Leif Eriksson, his son, went even farther: casting off from Greenland, he sailed nearly 2000 miles (3219 km.) to a land he named Vinland, a part of the area which we now call North America. The ships that carried these expeditions, called *kanerrirr* in the

Norwegian tongue, had broad beams to keep them from rolling and could carry 20 tons and up to 15 people. Indeed, just after the year 1000, Thorfinn Karlsefni led a flotilla of such ships carrying cattle and between 65 and 165 colonists from Greenland to Vinland. Although the colony ultimately failed, the courage of Thorfinn Karlsefni is memorialized along with that of Erik the Red and Leif Eriksson in sagas repeated from generation to generation.

The nautical achievements of the Norse, however spectacular, were infrequent. But voyages of such distances were so commonplace among the Polynesians as scarcely to merit comment. Polynesians thought little of sailing from Fiji to Tonga (422 miles [679 km.]) or Samoa (769 miles [1238 km.]), or from Samoa to Tahiti (1059 miles [1705 km.]). Admittedly, 3200-kilometer voyages comparable to Leif Eriksson's were uncommon in Polynesia, but even so, the 2700 miles (4348 km.) that separate Tahiti and Hawaii were traversed frequently enough to populate the entire archipelago, and the precise star course has been preserved in chants.

Polynesian oceangoing vessels were of various designs, but the finest were made in Fiji. "The Feejee canoes are superior to those of other islands," reported Charles Wilkes, commander of the U.S. Exploring Expedition of 1838–42.

> They are generally built double, and those of the largest size are as much as one hundred feet in length. . . . The sails are so large as to appear out of all proportion to the vessel, and are made of tough yet pliable mats. . . . It is the custom of the chief always to hold the sheet; thus it is his task to prevent the danger of upsetting. They steer with an oar having a large blade. In smooth water these canoes sail with great swiftness, but from the weight and force of the sail they are much strained, leaking at times very badly, requiring always one and sometimes two men to be constantly bailing out the water. Notwithstanding all this, they make very long voyages,—to Tonga, Rotuma, and the Samoan islands. The canoes are generally built of the vas [*vesi*] wood.

If the finest oceangoing vessels in the Pacific were made in Fiji, the finest ships in Fiji traditionally were made on tiny Kabara (Kahm-*bah*-rah) in the Lau group of islands, approximately 240 kilometers southeast of Suva. Commissioning an oceangoing raft in Kabara is not a simple matter. Any potential purchaser must surmount three challenges. First, the purchaser must choose from several types of sea craft. The *camakau* (thah-mah-*cow*) is a single-hulled canoe up to 15 meters in length used for interisland transport and warfare. The *drua* (*ndroo*-ah) has two hulls and requires up to 50 men to sail it. The *tabetebete* (tahm-bay-tay-*bay*-tay) is the largest of all Fijian sea craft, with an intricate hull formed of carefully fitted planks. One measured in 1860 by the missionary Thomas Williams was found to be 36 meters and 7.3 meters. A unique perspective on the *tabetebete* was provided by a sandalwood trader, William Lockerby, who had been held captive in its brig:

> Round the sides of the platform there is a strong breastwork of bamboos, behind which [the warriors] stand in engaging an enemy. There is also a house on a platform which is erected and taken down as circumstances require. The number of men on board amounted to two hundred. Captain Cook's account of the swift sailing of these vessels is quite correct, however incredible it may appear to those who have not seen them. With a moderate wind they will sail twenty miles an hour.

A sailing vessel capable of transporting 200 to 300 warriors across the open ocean at 20 miles an hour represents the indigenous equivalent of a C5 military

The Polynesian islands, including the island of Kabara where the finest sea craft were made, are contained within a triangle defined by vertices at Easter Island, New Zealand, and Hawaii. The Polynesians made long voyages throughout this area, and perhaps beyond. New Zealand was colonized in epic migrations from the Society Islands, which followed the route of the Maori canoe *Aotea*. The sweet potato, *Ipomoea batatas*, entered Polynesia from South America.

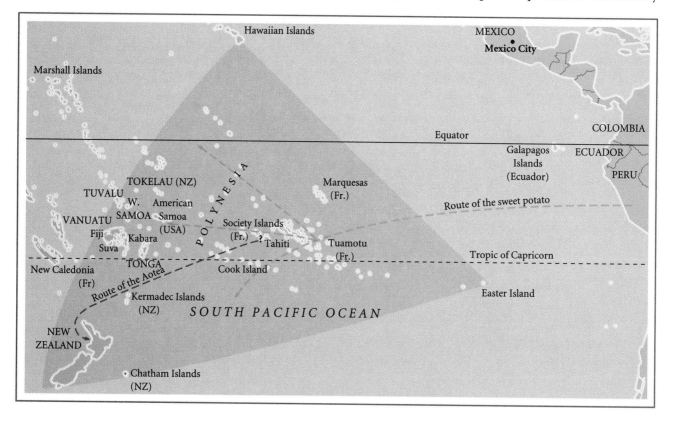

aircraft, a formidable military platform capable of projecting raw power into distant conflicts. Possession of such a craft could easily decide any interisland dispute. Thus the shipwrights of Kabara, in their heyday, before contact with Europeans in the eighteenth century, functioned as a combination Boeing and Lockheed for the entire South Pacific.

Once the type of craft has been selected, a second obstacle confronts the potential purchaser of a Kabaran vessel: the purchaser must satisfy the strict dictates of Kabaran custom in the ordering process. A whale tooth, or *tabua*, for the island chief must accompany any order to certify that the client is economically powerful and politically important. After the chief has received his *tabua*, he must consult with the village elders and give them a second whale tooth supplied by the purchaser. The elders' approval cannot be taken for granted, for possession of a Kabaran sea vessel has historically been the decisive factor in many interisland conflicts. As naval armorers for clients from the distant king of Tonga to the nearby king of Lakemba, Kabarans, like all good arms merchants, have needed to maintain strict neutrality in conflicts, or at the very least to be careful never to give serious offense to an ascendant military power. Tiny Kabara could never have withstood a sustained siege: although it is considered a "cave of wealth" because of its unique hardwood forest, it is also known as a "famine island" because of its dearth of arable soil. There is no room for error in the political calculations of the village elders.

If the elders agree, they must offer a whale tooth to one of two resident shipwright guilds and ask them to build the vessel. This request, too, is not a mere formality; the shipwrights reserve the right to withhold their extraordinary expertise. The Kabara shipwrights are the best of the best, the finest to have ever lived in the South Pacific. A strike by either guild would stop work on all sea craft under construction. A serious offense to the guild might even provoke sabotage by magic. A single wedge of *matalafi* (*Psychotria insularum* [Rubiaceae]) hidden under the lashings of the bow is a curse that can doom any vessel. If the guild is properly approached, however, its approval is almost certain, because construction of a single large vessel can ensure the shipwrights prosperity and good food for as long as three years.

Should a Kabaran oceangoing craft be commissioned today, a third and more serious difficulty would have to be surmounted: traditional knowledge concerning which plants to use for construction has been largely forgotten. The esoteric knowledge of the shipwright guild vanished almost entirely in the wake of the European colonization of Fiji in the nineteenth century. Indigenous shipwrights had begun to disappear from Fiji by the turn of the century, as had indigenous nautical expertise throughout the Pacific. Even the famous *Hokule'a*,

Only a few older craftsmen such as Josafata Cama, a shipwright resident in Kabara Island, retain the skill of constructing large Polynesian seacraft. The advent of modern tools and a renewed interest in Polynesian voyaging may succeed in saving the remarkable skill of these expert craftsmen.

guided in the 1970s by one of the few remaining traditional navigators, Mau Pilag of the Caroline Islands, was a fiberglass re-creation of a traditional Polynesian design.

Imagine our surprise when one of us learned in a phone call that indigenous nautical design was still possible in Kabara. Jerry Loveland, director of the Institute for Polynesian Studies, reported to Paul Cox that an aged shipwright, Ilaijia Ledua, remembered how to build oceangoing craft. He was not sure that he had the resources to build a *drua,* but he offered to direct the construction of a single-hulled canoe, a *camakau.*

Three months, and several whale teeth later, Cox received an aerogram from Sandra Banack, one of his graduate students, working in Kabara. Cox was concerned that the Kabara islanders might not accept a woman as an investigator. Written accounts of Fijian canoe manufacture reported that women were not even allowed to approach the construction site, let alone participate in the manufacture of a boat. Menstruating women, in particular, were believed to bring disaster on the maiden voyage. Cox had instructed Banack to "tell the shipwrights that as a scientist you enjoy all the rights of men in our culture, and that for all cultural purposes, you function as a man."

"The village elders say that there is no problem with my documenting the construction," a delighted Banack reported, "since I am viewed as the representative of the commissioning party, rather than as a woman. I am the only woman allowed near the boat, but the carpenters allow me complete access and let me photograph everything."

Sandra Banack and her husband David had arrived on Kabara after a rough two-day sail from Suva, bringing with them a mountain of video gear, plant presses, and the determination to photograph and collect every plant used in the construction of the 15-meter *camakau*. The construction of this *camakau* would be historic—the first time a Polynesian oceangoing craft had been entirely fabricated in the presence of a trained ethnobotanist. After a three-month crash course in Bauan Fijian taught by a native speaker at Brigham Young University, the Banacks left to take up residence on Kabara for six months.

Sandra Banack sought to discover why Kabara was so prominent in ancient Polynesian politics. Why had that tiny island attracted the finest shipwrights of the Pacific? (One Tongan shipbuilding clan had been resident there for generations.) Indeed, why did Kabara and the surrounding group of small islands in the Lau group become the most important political force in all of Fiji?

The answer became apparent shortly after Banack arrived. The first time she walked into the Kabara forest with Ilaijia Ledua, she found herself looking at the finest stand of *vesi* trees (*Intsia bijuga* [Caesalpiniaceae]) in the Pacific. The

Left: Tiny Kabara Island has little arable soil, but does contain a precious forest of *vesi* trees (*Intsia bijuga*). Here men search the forest for trees needed to build the hull of an oceangoing craft. Right: Since *I. bijuga* trees are extraordinarily heavy, islanders hollow out a tree in the forest before transporting it to the coastal construction site.

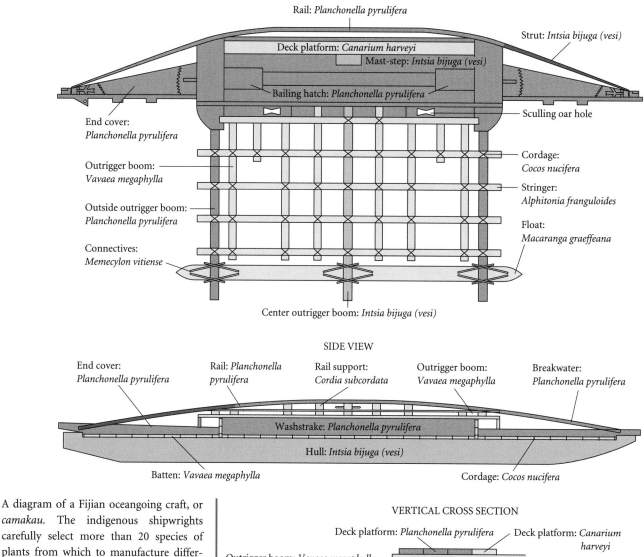

OVERHEAD VIEW

Rail: *Planchonella pyrulifera*

Deck platform: *Canarium harveyi*

Mast-step: *Intsia bijuga (vesi)*

Bailing hatch: *Planchonella pyrulifera*

Strut: *Intsia bijuga (vesi)*

End cover: *Planchonella pyrulifera*

Outrigger boom: *Vavaea megaphylla*

Outside outrigger boom: *Planchonella pyrulifera*

Connectives: *Memecylon vitiense*

Sculling oar hole

Cordage: *Cocos nucifera*

Stringer: *Alphitonia franguloides*

Float: *Macaranga graeffeana*

Center outrigger boom: *Intsia bijuga (vesi)*

SIDE VIEW

End cover: *Planchonella pyrulifera*

Rail: *Planchonella pyrulifera*

Rail support: *Cordia subcordata*

Outrigger boom: *Vavaea megaphylla*

Breakwater: *Planchonella pyrulifera*

Washstrake: *Planchonella pyrulifera*

Hull: *Intsia bijuga (vesi)*

Batten: *Vavaea megaphylla*

Cordage: *Cocos nucifera*

A diagram of a Fijian oceangoing craft, or *camakau*. The indigenous shipwrights carefully select more than 20 species of plants from which to manufacture different parts of the canoe. They use *Intsia bijuga*, because of its extraordinary tensile strength, for parts of the *camakau* likely to be placed under high stress, such as the hull, the center outrigger boom, and the seat for the mast. Lighter woods, not as prone to sink, are used for other parts of the canoe.

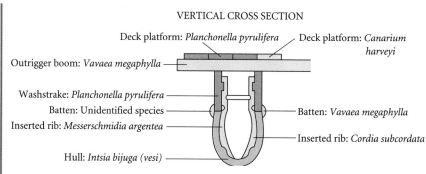

VERTICAL CROSS SECTION

Deck platform: *Planchonella pyrulifera*

Deck platform: *Canarium harveyi*

Outrigger boom: *Vavaea megaphylla*

Washstrake: *Planchonella pyrulifera*

Batten: Unidentified species

Inserted rib: *Messerschmidia argentea*

Hull: *Intsia bijuga (vesi)*

Batten: *Vavaea megaphylla*

Inserted rib: *Cordia subcordata*

importance of these trees to the islanders became clear when the head carpenter intoned a lengthy prayer before allowing any of the young men to touch an ax to the tree he had selected for the hull. "The Vesi is one of the sacred trees of Viti," Berthold Seemann wrote in 1865.

> Europeans have occasionally come in unpleasant contact with the Fijians, when unknowingly they had cut it down for timber. . . . The wood is the best in the islands, and seems almost indestructible; it is used for canoes, pillows, kava bowls, clubs, and a variety of other articles.

Vesi trunks are the Polynesian equivalent of structural steel, the only plant material that can withstand the tremendous stresses encountered by a large ship's hull in heavy seas. Although the *vesi* is widespread, ranging eastward from Indo Malaysia to Tonga and Samoa, it grows extraordinarily tall on the tiny upraised limestone platform of Kabara. From Kabaran *vesi* it is possible to fashion a single-piece hull that exceeds 20 meters in length.

The new *camakau* was to be 15 meters long. After the prayer, Ilaijia Ledua directed the young men to fell the tree. They cleared the trunk of branches and hollowed it before they dragged it over rollers to the construction site on the shore. Then they harvested other *vesi* trees for the masthead, steering oar, and other parts of the *camakau* where tensile strength and durability would be critical.

Vesi's only drawback as construction material for a boat is its remarkable density. Since the weight of all the wood used to construct a ship must not exceed the buoyancy of the vessel, less dense woods must be used for its other parts. The shipwright chose a species that grows in gaps in the rain forest, *Macaranga graeffeana* [Euphorbiaceae], for the float because of its light, porous wood. Banack recorded more than 20 species of indigenous plants used in the construction of the canoe. Ilaijia Ledua functioned like a skilled aeronautical engineer, carefully choosing each plant on the basis of its performance characteristics. Plants used to plait cords and ropes had to be selected for strength and durability. Sennit lashings, woven from coconut husks, were carefully caulked with the sap of *Canarium harveyi* [Burseraceae], which afforded them some protection from seawater. The rigging ropes, however, could not be similarly protected and so were woven from *Hibiscus tiliaceus* [Malvaceae], a coastal shrub. Hibiscus fibers are more difficult to plait than coconut fibers but are much longer, so cordage made from them has superior shear strength.

The 30-square-meter sail was woven from *Pandanus* [Pandanaceae] leaves by the village women. Traditionally the women stitched the pieces together with

There are ritual prohibitions against women working on a *camakau* except for the sail, which Vika Usu and other women weave entirely from *Pandanus* leaves.

The maiden voyage of a Fijian *camakau* off the coast of Kabara Island. Visible on the prow of the canoe is ethnobotanist Sandra Banack.

a needle carved from the shinbone of a slain enemy, but for our sail they used a commercial sail needle.

The day arrived when the vessel was completed, and a ceremonial feast was prepared for the shipwrights. A launching that morning seemed unlikely, however. The seas, churned by two-meter swells, were the roughest Banack had seen during her six-month sojourn on Kabara. The villagers feared that the craft might capsize if it were launched. Yet as a special film crew loaded movie cameras in their motor launch, Banack was startled to see a group of village men approach the ship and begin to push it off the beach into the sea. Banack ran to ask what was happening.

"We are leaving," the crew told Banack. "Jump in!"

They had no life jackets and no charts, and, as Banack knew, the carpenters had no experience as sailors. Although Ilaijia Ledua had managed to preserve the knowledge of the shipwright guild, the navigator's guild had long since vanished. Only one living Kabaran had ever sailed a *camakau*. Banack faced a crucial decision, but her training as an ethnobotanist to stay with her teachers paid off: she jumped in the boat. As the wind filled the large *Pandanus* sails, the *camakau* took off like a rocket, leaving the motor launch far behind.

Polynesian Voyaging and the Discovery of New Zealand

Perhaps it was massive sea craft similar to the *camakau* that originally brought the ancestors of the Maori to New Zealand. Roger Green, an archaeologist at the University of Auckland, believes that the first Maoris brought with them a very early form of eastern Polynesian culture, which they adapted to a land rich in new resources: thousands of flightless moa birds, some standing twice the height of a man; an inexhaustible supply of the *aruhe,* or bracken fern, *Pteridium aquilinum* [Dennstaedtiaceae], whose roots supplied a staple of the Maori diet; and fertile soil and a climate ideal for cultivation of the sweet potato. The date of original discovery is unclear; although archeologist Janet Davidson suggest that by A.D. 1200 the coast of New Zealand had been settled, other archeologists posit that New Zealand may have been discovered as early as A.D. 800. The nature of colonization, whether a single event or episodic, is also difficult to interpret from archeological evidence. But Maori legends point to at least one colonization event that is without parallel elsewhere in the world: a single diaspora of several thousand people reaching North Island as a flotilla.

Maori legends are replete with stories of the canoes that made the original voyage. Indeed, there is scarcely a Maori alive today who cannot trace his or her ancestry to the arrival of a specific *waka,* or canoe, in the great migration from Havaiki, the ancestral island home of the Maori, whose location is shrouded in the mist of time but is probably near Tahiti. Ethnohistorical accounts place this migration in the early fourteenth century, at roughly the same time that the Anasazi people were completing their mass migration from the Colorado Plateau to the southern mesas. Tree ring datings make it clear that the Anasazi migrated when prolonged drought created famine, but what was the impetus for the Maori migration?

The fragile ecology of oceanic islands allows little tolerence for mismanagement. We do not know what happened in Havaiki, but we do know that the interior of the island of Mangareva became almost useless for agriculture as a result of deforestation. Similarly, the scrubby rainforests of Easter Island were almost completely destroyed, dramatically increasing soil erosion and, as archeologist Patrick McCoy argues, producing a shortage of wood for canoes. This in turn likely reduced fishing yields. In Tikopia (a Polynesian outlier in the Solomon Islands) Patrick Kirch and Douglas Yen found that pigs suddenly van-

Song of an Aotea Oarsman

Fiercely plies the shaft of this my paddle
Named *Kautu ki te rangi.**
To the heavens raise it,
To the skies uplift it.
It guides to the distant horizon,
To the horizon that seems to draw near,
To the horizon that instills fear,
To the horizon that causes dread,
The horizon of unknown power,
Bounded by sacred restrictions.
Along this unknown course
Our ship must brave the waves below,
Our ship must fight the storms above.
This course must be followed.

* "Directed to the heavens."

ished from the archeological record about 200 years ago. Whether the pigs succumbed to disease, gluttony, or unknown cultural changes, the consequence was the same: the disappearance of the only domestic mammal on the island resulted in the loss of an important source of protein.

Although we may never know all the ecological precursors to the Maori migration, in a startling parallel to the Anasazi migration almost all of the Maori canoe traditions suggest that fighting over dwindling food supplies in Havaiki led to the mass flight. Just as the consciousness of the Jewish people has been molded over centuries by stories of their exodus from Egypt to a promised land "flowing with milk and honey," the culture of the Maori has been shaped by oral histories of the mass migration to Aotearoa, "land of the long white cloud," the Maori name for New Zealand.

The exodus of the Maori from Havaiki to Aotearoa ranks as one of the greatest migrations in human history. Braving wind, wave, and unknown peril, thousands of people set out to cross over 2000 nautical miles (3706 km.) of open ocean to New Zealand. The stories of this epic migration, like the Norse sagas of Eric the Red, have been related from generation to generation. The ancient Maori poem reprinted in the margin, as translated by Te Rangi Hiroa, poignantly captures the feelings of an oarsman on the canoe *Aotea* as he set sail.

Sometime after the *Aotea* departed Havaiki with its companion canoe the *Ririno,* disputes over navigation threatened the voyage. Potoru, captain of the *Ririno,* said that the canoe should set a course for the setting sun, but Captain Turi of the *Aotea* insisted that the course set by his navigator Kupe, be followed precisely, and he directed the bow toward the rising sun. The canoes thus separated and the *Ririno* was lost. To this day the Maori say of a stubborn person, "He persists in the obstinacy of Potoru." Legend claims that the *Aotea* landed on Rangitahua in the Kermadec Islands to relash the top strakes of the canoe and offer sacrifices to the gods. Having made the necessary repairs, the crew again set sail for Aotearoa.

Obviously such lengthy voyages could not be undertaken in small craft. Maori legends indicate that most of the migration canoes, single-hulled like the Fijian *camakau,* were of enormous length. Most of the other canoes landed on the east coast of North Island, but the *Aotea* landed on the west coast, near the present town of Kawhia. The migrants planted crops as soon as they landed, but met with mixed success. Breadfruit and coconuts, those staples of the Polynesian diet, could not survive the cool climate of New Zealand. Legends of failed coconut plantings record what must have been a crushing blow to the new colonists: "*Ni* was the name of that fruit, the size being about that of a child's head; that kind of fruit was brought here . . . but they never grew."

The crew of the *Aotea*, however, had carried with them a most precious cargo. Their priest had secreted nine seed sweet potatoes in his belt. He sacrificed one seed potato as an offering, but the other eight potatoes were immediately planted.

Sweet Potatoes and the *Kon-Tiki* Expedition

We might regard the Maori sagas of the exploration and colonization of New Zealand as unverifiable claims of descent from heroic voyagers, except for those nine seed sweet potatoes. This detail, together with stories of sweet potatoes brought on other canoes, is potentially subject to experimental verification.

In 1947 the Norwegian adventurer Thor Heyerdahl suggested that the presence of sweet potatoes in both Peru and the islands of Polynesia provided strong evidence that Polynesia had been colonized by American Indians:

> The sweet potato [that] Tiki [an ancient god] brought with him to the islands, *Ipomoea batatas,* is exactly the same as that which the Indians cultivated in Peru from the oldest times. Dried sweet potatoes were the most important travel provisions both for the seafarers of Polynesia and for the natives in old Peru. In the South Sea islands the sweet potato will grow only if carefully tended by man, and, as it cannot withstand sea water, it is idle to explain its wide distribution over these scattered islands by declaring that it could have drifted 4,000 sea miles with ocean currents from Peru. This attempt to explain away so important a clue to the Polynesians' origin is particularly futile seeing that philologists have pointed out that on all the widely scattered South Sea islands the name of the sweet potato is *kumara,* and *kumara* is just what the sweet potato was called among the old Indians in Peru. The name followed the plant across the sea.

To test this hypothesis, Heyerdahl built a raft of Peruvian balsa wood (*Ochroma lagopus* [Bombacaceae]). He modeled the craft, which he named the *Kon-Tiki,* after early European drawings of South American Indian oceangoing craft. In an epic voyage of 101 days from Lima to the Tuamotu archipelago, south of Tahiti, Heyerdahl proved that a prehistoric crossing from South America to Polynesia in a primitive craft and with minimal navigational aids was possible. But mere possibility is a far cry from plausibility. Though scholars admired Heyerdahl's sense of adventure, they viewed his conclusions with suspicion. Some argued that there were sweet potatoes in Polynesia before the Europeans colonized the islands, having been brought there by early Spanish

The sweet potato *Ipomoea batatas* is a staple throughout Polynesia. Ethnobotanist Douglas Yen has amassed genetic and linguistic evidence that strongly points to a South American origin for the tuber.

Indigenous Names for Sweet Potatoes (*Ipomoea batatas*)

South America	
Urubamba, Peru	*cumara*
Cuzco, Peru	*cumara*
Sierras, Peru	*cjumara*
Lima, Peru	*kumar*
Ecuadorian Highlands	*cumar*
Colombia	*kuala*

Polynesia	
Easter Island	*kumara*
Tuamotu	*kumara*
Mangareva	*kumara*
Rarotonga	*kumara*
Marquesas	*kuma'a*
Aitutaki	*kuara*
Tahiti	*umara*
Samoa	*umala*
Tonga	*kumala*
Hawaii	*'uala*
Niue	*timala*
New Zealand	*kumara*

explorers. Others argued that seeds might have been dispersed to the islands by ocean currents.

In a study of sweet potatoes throughout South America, Polynesia, Melanesia, and Southeast Asia, the ethnobotanist Douglas Yen of the Bishop Museum in Honolulu determined that the center of diversity for sweet potatoes is South America on the basis of four general pieces of evidence. Studying a variety of morphological characters, such as leaf shape, hairiness of stems, and root shape, Yen found South American sweet potato populations to be far more diverse than other populations. He also found archaeological remains of sweet potatoes in South America to be of very ancient date. Polynesian legends, too, all seem to indicate that the crop originated in the east. Linguistic evidence is also consistent with a Peruvian origin.

These as well as genetic considerations led Yen to conclude that the ancestry of all Pacific sweet potatoes could be traced to South America. Resolving the conundrum of the sweet potato's origins does not substantiate Heyerdahl's hypothesis, however, since prehistoric voyaging is only one possible mechanism of transfer. "It may be necessary to revert to an earlier question—whether indeed there were any direct prehistoric contacts between America and Polynesia," Yen wrote. "The plant evidence must remain inscrutable on the subject, for it can never make a positive identification of the transferrer. At best it gives direction."

More recently the linguist Karl Rensch of the Australian National University has been intrigued by the Hawaiian name *'uala*. Using generalized rules for patterns of consonant loss in Polynesian languages, Rensch finds *'uala* an unlikely derivative of the Polynesian *kumara*, since the change of *m* to no sound at all does not occur in any other Polynesian language. Rensch suggests instead a linkage with the Cuna language family of northern Colombia.

Despite the considerable evidence for dispersal of sweet potatoes from South America to Polynesia, few scholars believe that Polynesia was populated from South America. Blood type groupings, linguistics, archaeological evidence, and studies of indigenous agriculture all suggest that the Polynesian people descended from the Lapita, an agricultural people who long ago left Indo Malaysia and began a migration that eventually took them to the islands of the South Pacific. There may, however, have been limited prehistoric contact between South America and Polynesia, and some Polynesian groups may have obtained sweet potatoes that way. Certainly once sweet potatoes were introduced, they could have been rapidly distributed throughout Oceania in the superb Polynesian sailing vessels.

Polynesian voyaging had an immense impact on the patterns of colonization and human settlement in the Pacific islands. Those impacts have been studied by a variety of anthropologists, linguists, archaeologists, and biogeographers, but investigators reared in industrial societies often forget the mundane details of ship construction—the means used to attach one part to another, for example. With no epoxy resins or steel cable at hand, preindustrial societies needed to rely on plants to produce adhesives and cordage. Ethnobotanists have focused attention on cords, glues, fasteners, and containers—things that are ubiquitous in modern life but had to be carefully constructed in nonindustrial settings, not just in Polynesia but in all indigenous societies, including those of temperate North America.

Cordage and Containers

Using spruce adhesive and basswood twine, the Chippewa Indians were able to create an astonishing variety of implements from their primary building material: birch bark. In birch bark canoes, stitched with basswood twine and caulked with spruce gum, the Chippewa could go on fishing expeditions and collect wild rice. They even used large sheets of birch bark, stitched together with basswood twine, to cover their dwellings.

To make the spruce gum adhesive, the Chippewa boiled *Picea rubra* [Pinaceae] gum in a mesh bag, skimmed the resin from the surface, and then mixed it with cedar charcoal. They made their cords from basswood bark (*Tilia americana* [Tiliaceae]), which they cut in long strips and then soaked in water for several days to allow the inner bark to be separated from the outer bark.

The Chippewa's fascination with birch bark was not merely utilitarian; the bark had deep religious significance. "As long as the world stands this tree will be a protection and benefit to the human race," said the deity Winabojo, who had hidden in a birch tree to protect himself from thunderbirds. "If they want to preserve anything they must wrap it in birch bark and it will not decay." In accordance with this promise, the Chippewa found they could store maple syrup for over a year in birch bark satchels, some of which held as much as 9 kilograms. The Sami people in Lapland also used birch bark for containers, and they still use bowls carved from birch wood to collect reindeer milk.

The use of plants as containers was even more crucial among cultures that needed to transport water over long distances. Because of political disputes or the need to defend themselves in the event of attack, many people choose to live at some distance from a water supply. South Americans used large bottle gourds of the genus *Lagenaria* [Cucurbitaceae] to transport water. Similar gourds have been found in the Pacific islands; some gourds found in Hawaii could hold

This reindeer milking bowl, or kåsa, was used by the Sami people of Lapland. Made of birch wood (*Betula pendula* [Betulaceae]) with an engraved reindeer bone handle, it is designed to be held in one hand, leaving the other hand free to milk the reindeer. Birch wood was used not only by the Sami people, but also by many Native American tribes such as the Chippewa to make containers for preserving food.

The Maori used not only bull kelp, but gourds like this one, called a *taha huahua,* to preserve muttonbird. The gourd was often fitted with a cover woven from New Zealand flax (*Phormium tenax* [Agavaceae]) and ornamented with bird feathers. A lid carved from *Prumnopitys taxifolia* [Podocarpaceae] was used as a stopper. Cooked flesh of birds packed in this manner could be preserved for two years or more.

more than 10 liters of liquid. Plants can also be used as containers to preserve food.

The Maori fashioned one ingenious type of container from hollow pieces of the bull kelp, *Durvillaea antarctica* [Durvillaeaceae]. They made a small hole in the side of the kelp and then inflated and dried the entire kelp. When the kelp was dry, it retained the inflated shape and resembled a huge bottle. Not only oil was stored in seaweeds, but also, astonishingly, the flesh of muttonbirds of the genus *Puffinus* [Procellarridae]. By placing these small birds in the seaweed pockets and pouring the birds' own melted fat on top of them, the Maori created a germproof seal, in much the same way we produce canned beef today. As in any modern meat-packing plant, strict taboos, or *noa,* were imposed on the workers. Workers were forbidden to eat any of the food they were packing, and they constantly recited special charms, or *karakia.* Such care was important not only because of Maori religious proscriptions but also because of the potentially fatal consequences of botulism poisoning.

Still, it is difficult to compare Maori methods of aseptic preserving with modern industrial canning processes based on pressure cookers and constant bacteriological monitoring. Some indigenous technologies, though, can indeed be compared in effectiveness to their modern equivalents. With the exception of technological innovators in China, no indigenous peoples are known to have developed anything remotely resembling gunpowder, but adroit uses of plants made it possible for many of them to enjoy fishing or hunting successes comparable to the yields obtained with modern firearms.

Arrow Poisons

With the exception of the Australian Aborigines, continental peoples throughout the world have used arrow poisons for hunting and warfare. The indigenous peoples of South America used poisoned arrows primarily to hunt birds and monkeys, but the Africans took far larger game. In 1861, during David

Arrow poisons of remarkable toxicity have been developed throughout the world, although the plants used vary from continent to continent.

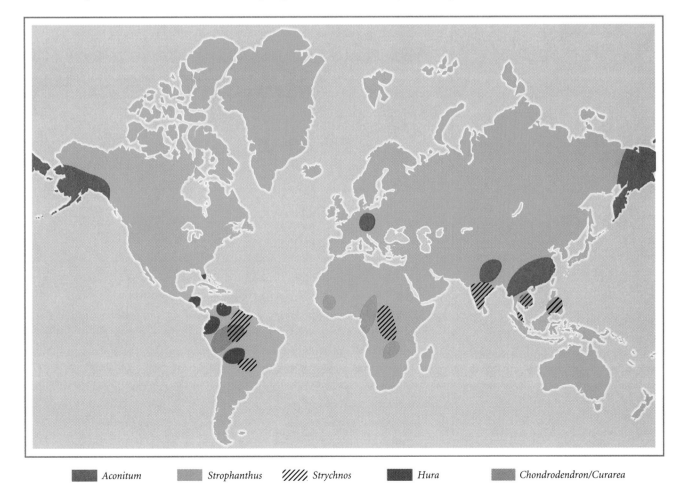

| Aconitum | Strophanthus | //// Strychnos | Hura | Chondrodendron/Curarea |

Livingstone's exploration of the Zambesi, the botanist John Kirk attempted to discover the source of a mysterious poison called *kombé,* powerful enough to down a Cape buffalo. "The arrow making no noise, the herd is followed up until the poison takes effect and the animal falls out," wrote Livingstone. "It is then patiently watched til it drops—a portion of meat round the wound is cut away, and all the rest eaten. . . . It is possible that the *Kombi* may turn out a valuable remedy."

Even larger game, such as hippopotamus, could be taken if the poison could be made to penetrate the tough hide. Kirk noted that

> the hippopotamus is killed by it, but the quantity needed seems to be thrice that of an ordinary arrow. It is driven through the thick skin of the animal by being placed on the barbed head in the lower end of a beam of wood, which falls from a height as the beast passes underneath a trap. The poisoned head is driven well in by the big end of the beam, and is left to act, which it is said to do in about half a day.

The indigenous hunters concealed the source of *kombé* from Kirk. "I had long sought for it, but the natives invariably gave me some false plant," he wrote in a letter to Sir Thomas Fraser, a pharmacologist at the University of Edinburgh.

Indigenous peoples on several continents have discovered how to make potent arrow poisons from plants. In the Zambesi River region of Africa, arrows tipped with poison from *kombé (Strophanthus kombe)* could down animals as large as a hippo; the cardiac glycoside G-strophanthin derived from the plant is now used as a remedy for congestive heart failure. Tubocurarine, an alkaloid derived from the South American plant *Chondrodendron tomentosum,* is placed on the tips of blow darts; it is used medicinally as a muscle relaxant during surgery. Strychnine, an alkaloid obtained from the seeds of *Strychnos nux-vomica,* found in India and Sri Lanka, is seldom employed in medicine but is used in neuroanatomical research. The monk's hood genus, *Aconitum,* produces the alkaloid aconitine. *Aconitum* species were used in China and Europe as arrow poisons.

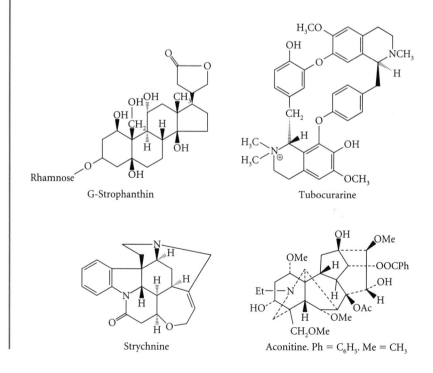

G-Strophanthin

Tubocurarine

Strychnine

Aconitine. Ph = C_6H_5. Me = CH_3

The Waorani Indians of Ecuador prepare curare for poison darts by scraping the bark of *Chondrodendron* vines (left), then filtering the bark through palm leaves to extract their lethal compounds. In the hands of Waorani, curare dramatically increases hunter efficiency, giving yields approaching those obtained by shotguns.

One day at Chibisa'a village, on the river Shiré, I saw the *Kombé*, then new to me as an East African plant. There climbing on a tall tree it was in pod, and I could get no one to go up and pick specimens. On mounting the tree myself to reach the *Kombé* pods, the natives, afraid that I might poison myself if I handled the plant roughly or got the juice in a cut or in my mouth, warned me to be careful, and admitted that this was the '*Kombé*' or poison plant. In this way the poison was identified, and I brought specimens home to Kew, where they were described.

Kirk accidentally contaminated his toothbrush with the plant and noticed that his pulse slowed as soon as he brushed his teeth. He sent samples to the

The toxicity of various arrow poisons developed by indigenous peoples is plotted as the inverse of the LD50, the dose required to kill 50 percent of animals in a trial. The vertical axis is logarithmic, thus Huratoxin manufactured from *Hura crepitans* is 500,000 times more toxic than cyanide. Such highly toxic substances allowed indigenous hunters equipped with darts or arrows to achieve yields comparable to those obtainable by shotguns.

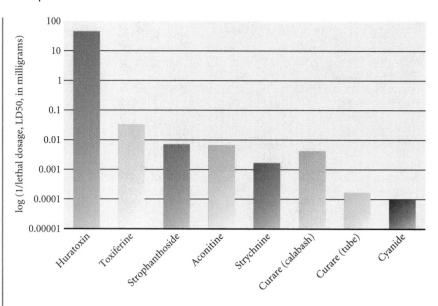

Royal Botanic Gardens at Kew, where the plant was named as *Strophanthus kombe* [Apocynaceae]. In an early interdisciplinary collaboration, Kirk and Livingstone joined forces with Thomas Fraser, a pharmacologist at the University of Edinburgh, to study *kombé*. John Buchanan, British consul in the Nyassa district, described the preparation of the poison:

> A man breaks a [*Strophanthus*], fruit, and puts the seeds with wool attached into a pot. He then takes a small piece of Bamboo, which has two thin splints inserted crosswise in the ends, and he revolves this speedily by rubbing it between his hands. The seeds are thus put into motion and fall to the bottom of the pot, and the wool rises and comes out at the top, and is carried away by the least breath of the wind. The seeds are then put into a small mortar and pounded into a paste, which is then ready for use. It is common to mix the milky juice of a *Euphorbia* with it to make it stick on the arrow.

Through careful pharmacological work, Fraser confirmed that the active substances in *Strophanthus* have a profound effect on the heart. Using bioassay-guided fractionation (through experiments on frogs and rabbits), Fraser isolated a glycoside known as strophanthin, a drug that is now used to treat heart failure.

South American arrow poisons also yield therapeutically useful drugs, particularly curare, used by tribes that hunt in the tropical rain forests. From extensive pharmacological studies of curare, Norman Bisset at King's College has found that curare carried in calabashes is typically made from species of *Strychnos* [Loganiaceae], whereas curare carried in bamboo tubes is made from *Chondrodendron* [Menispermaceae] or *Curarea* [Menispermaceae]. Unlike the African *Strophanthus* poisons, curare does not affect the heart but instead is a muscle relaxant that kills by paralyzing the muscles required to breathe. Tubocurarine, so named because it was isolated from curare carried in bamboo tubes, and toxiferine, isolated from curare carried in calabashes, have both become crucial anaesthetic drugs for use in surgery; some surgeries, particularly open-heart surgery, would be impossible without these compounds or synthetically modified derivatives.

How effective are the arrow poisons in native settings? In a study of the Waorani people of Ecuador, James Yost and Patricia Kelley compared the effectiveness of a blowgun, a poisoned arrow, and a poisoned spear with that of a shotgun. They found that hunters obtained yields only 22 percent greater with a shotgun than with poisoned arrows or spears. The efficiency of blowgun poisons is clearly due to their tremendous potency. The huratoxin in arrow poisons manufactured in the Caribbean from *Hura crepitans* [Euphorbiaceae] was half a million times more toxic than potassium cyanide. Strophanthoside from *Strophanthus* is 77 times more toxic, and even crude curare from the calabash is five times more toxic than cyanide.

Arrow poisons ultimately represent a calculated response to the material contingencies of life: wild game is an abundant source of the protein needed for survival. That people should use plants to fill pressing needs for food, shelter, and transportation is not unexpected. What is extraordinary is the amount of effort indigenous peoples devote to plant uses that are not essential for their survival. Even in modern societies, the strength, attractiveness, and warmth of wood still make it preferred for fine cabinetry and furniture over more durable and cheaper plastics and metals. Such uses bespeak more than mere material constraints and provide some insight into human aesthetics and into the need for ways of expression that transcend the purely functional. Why else should indigenous (and modern) peoples produce tattoo inks, body paints, fragrant shampoos, or ornamented textiles? We find that such choices are not only a matter of taste but reveal much about the human condition.

The Body as Canvas: Plants and Tattoos

The beginning of May 1768 was an auspicious time for Samoa. During the previous two weeks Aotourou, the Tahitian navigator who accompanied Commander Louis de Bougainville, had done everything but commandeer the ship to set a course for these islands. Bougainville was stunned to learn that the Tahitians knew of other Polynesian peoples. "We had incontestable proof that the inhabitants of the isles of the Pacific Ocean communicate with each other, even at considerable distances," Bougainville wrote in his journal.

> The night was very fair, without a single cloud, and all the stars shone very bright. Aotourou, after attentively observing them, pointed at the bright star in Orion's shoulder, saying we should direct our course upon it, and that in two days' time we should find an abundant country. . . . As I did not alter my course, he repeated several times that there were cocoa-nuts, plantains, fowls, hogs, and above all, women, whom by many expressive gestures he described as very complaisant. Being vexed that these reasons did not make any impression upon me, he ran to get hold of the wheel of the helm, the use of which he had already found out, and endeavoured in spite of the helm's-man to change it, and steer directly upon the star which he pointed at. We had much ado to quiet him, and he was greatly vexed at our refusal. The next morning, by the break of day, he climbed up to the top of the mast, and stayed there all morning, always looking towards that part where the land lay, wither he intended to conduct us, as if he had any hopes of getting sight of it. He had likewise told us that night, without any hesitation, all the names that the bright stars we pointed at bear in his language.

About noon on May 3 the frigate *La Boudeuse* sailed through the strait between Ta'u and Olosega islands and was approached by Samoans in several dugout canoes. "They were naked, excepting their natural parts, and shewed us cocoa-nuts and roots," Bougainville wrote.

> Our Taiti-man stripped naked as they were, and spoke his language to them, but they did not understand him. . . . These islanders appeared to be of a middle size, but active and nimble. They paint their breast and their thighs, almost down to the knee, of a dark blue.

Had Bougainville been more observant, he would have noticed that the "dark blue" coloring of the Samoans was an indelible pigment inserted beneath the skin rather than on it. This was the first time Samoans had ever seen fair-skinned people, or *papalagi* (literally "burst from heaven"), and it was the first time Bougainville had ever seen a tattoo.

The blue pigment Bougainville observed is made from the nuts of the candlenut tree, *Aleurites moluccana* [Euphorbiaceae]. Called *lama* ("light") in

Samoan, the oily seed kernels of *A. moluccana* are threaded together on the midrib of a coconut leaflet and lit as a torch. To make the tattoo pigment, the Samoans gather *lama* nuts in coconut baskets and bake them in an underground oven for two days. They then crack the nuts with stones and thread them together as they do when they make a torch. They light the torch inside a special stone hearth. The torch emits a black, oily smoke, which settles as soot on the stone. The Samoans scrape the soot onto a banana leaf and then store it in a coconut shell.

The tattooing implements consist of a mortar and pestle to grind the pigment into a fine powder, several serrated combs of pig bone to pound the pigment into the skin, a mallet formed from a coconut leaf midrib to do the pounding, and a coconut shell palette. A towel made from bark cloth is used to wipe away the blood.

It typically takes between four and six weeks to receive a full Samoan body tattoo. The person receiving the tattoo lies on his stomach while the tattoo artist *(tufuga)* and his assistant sit on either side. The assistant stretches the client's skin tight. Working without a pattern, the tattoo artist dips the points of the comb in the pigment and starts tapping out a line. There is no room for error; the line must be absolutely straight and true the first time. After the artist has tapped more lines into the client's flesh, he inserts geometrical figures between them. Soon a highly stylized image of a canoe appears on the client's back. Filled areas appear as the serrations of *Pandanus* leaves, stylized bird heads, centipedes, and fishing nets. On the front the artist inserts two lines over either breast and a large triangle that stretches down to the pubis. Each thigh is successively tattooed, both front and back, down to the

Left: The soot obtained from burning the kernals of *Aleurites moluccana* makes a potent tattoo ink. The plant is also known as candlenut because the oily seeds can be burned for illumination. Right: Receiving a tattoo in Samoa is a prolonged and painful process, claimed to help men understand the travail of women during childbirth.

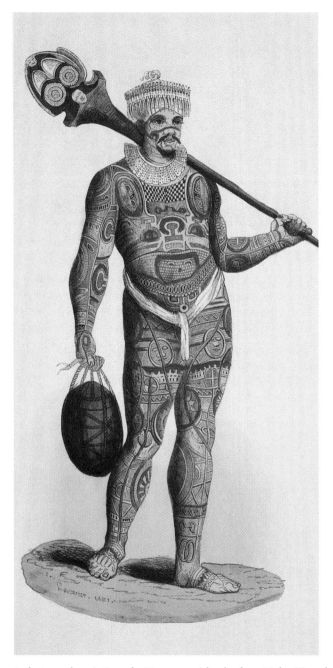 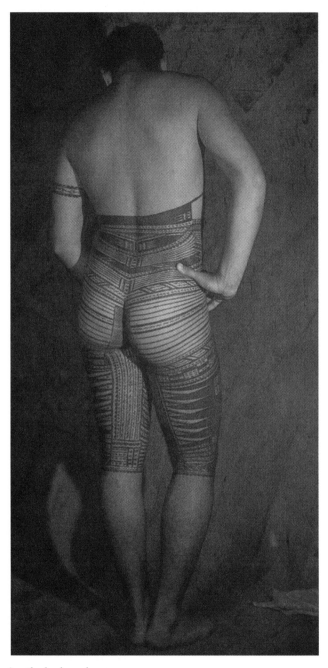

Left: An early painting of a Marquesan islander from Nuku Hiva, showing the body and facial tattoos. Right: Samoan tattoos, which cover the area between the knees and the breast, are composed of geometrical figures of mythical significance.

knee. The last tattoo is a short line over the navel. The navel mark certifies that the tattoo is complete.

Between the first mark on the back and the final navel mark on the front lies an arduous course of excruciating pain. Although most Samoan chiefs are tattooed, there is no shame attached to the lack of such body ornamentation. But woe betide the man who begins a course of tattooing and refuses to see it through to the end. He becomes *pe'a mutu,* a "tattoo coward," and is an object of scorn and ridicule for the rest of his days. Even worse, his family is humiliated for generations.

In view of the dire consequences of failure to complete a tattoo, why would a man start such a painful procedure? Tattooed Samoans admit that their motivations change during the course of the tattooing. At the beginning many men are motivated by vanity—a tattooed man is considered handsome—and a desire for increased acceptance in the village. Endurance of the pain also proves to any naysayer that one is remarkably tough and courageous. But these motivations fade with each successive tap, tap, tap of the *tufuga*'s mallet. Endurance to the end is motivated then not by vanity but by the enormous social consequences for the extended family if the tattoo is not finished. Samoans liken this change in motivation to pregnancy. An infant is conceived in pleasure, but as it continues to grow in the womb, the mother is motivated to bear her child not by pleasure but by love for her new family. A man who is not tattooed, Samoans believe, cannot understand the extended agony of childbirth or the intense love a mother develops for her child through her sacrifice. Tattooing, then, begins in a quest for personal pleasure but ends as a sacrifice of love for one's family. The link between tattooing and childbirth is made explicit in the traditional Samoan nuptials: the bride-to-be receives her dowry of fine mats and bark cloth, and the young man receives tattoos.

The art of indelibly ornamenting the human body was of enormous interest to the first European explorers in the South Pacific, who took several islanders back to Europe in order to exhibit them. Some explorers submitted themselves to the process. They called it "tattoo" in their efforts to approximate the Marquesan word *tatau.* Tricia Allen, an anthropologist at the University of Hawaii, reports that the first European to be tattooed was Jean-Baptiste Cabri, a sailor who jumped ship in the Marquesas in the late eighteenth century. His tattooed body was considered such a novelty in Europe that after his death his skin was preserved and publicly exhibited. John Rutherford, a resident of New Zealand, later received a Maori facial tattoo, or *moko.* Soon tattoos became a sign of a well-traveled sailor, and they remain so to this day.

Plants as Body Paints and Dyes

Just as the use of plant pigments for tattoos has cultural significance for Polynesians, so body paints derived from plants have meaning in cultures throughout the world. The Ka'apor of the Amazon Basin, for example, use a black pigment from *Licania heteromorpha* [Chrysobalanaceae] to paint the faces of their dead. The Berbers of North Africa also use black paints to color corpses. Unlike the Ka'apor paints, though, the Berber pigment is derived from henna, *Lawsonia inermis* [Lythraceae]. The ancient Egyptians also used henna to dye the black cloth in which they wrapped the dead.

Celts and Britons stained themselves bluish with woad, a dye made from the leaves of *Isatis tinctoria* [Brassicaceae], before engaging in battle. The word "Britain" is derived from *brith,* meaning "paint" in Celtic and "mottled" in Welsh. Woad-dyed warriors no longer confront visitors to Britain, but for many years the uniforms worn by British police officers were dyed blue with woad.

Kayapó children from Gorotire, a village in Pará state, Brazil, display traditional body adornment, applied before the celebration of a festival devoted to corn. The red color is derived from the seed arils (waxy coverings) of *Bixa orellana* [Bixaceae] and the dark purple from the fruit of *Genipa americana* [Rubiaceae]. This fruit has a clear juice that, when applied to the skin, shortly turns black and remains so for two weeks or more, gradually fading. The sap is sometimes mixed with charcoal that is used to paint body designs.

Body ornamentation can indicate more than status. The ethnobotanist William Balée of Tulane University found that Ka'apor women spend an average of 27 minutes a day working on objects to adorn themselves and their families. These decorations reflect not vanity but spiritual beliefs. Hardened resins from *Trattinnickia* species [Burseraceae] on a child's bead necklace can prevent serious illness; a piece of bark of *Petiveria alliacea* [Phytolaccaceae] can ward off evil spirits.

In view of the cultural potency of bodily decoration, it is no surprise that most peoples devote a great deal of time and effort to enhancing their personal appearance. The highlanders of New Guinea will even call off a war if they fear that rain will ruin their hair feathers. In Western societies, the cosmetics industry reaps billions of dollars in annual profits. Clothing, too, can serve functions that transcend the merely utilitarian. Rare is the army without uniforms, the church without priestly vestments, or the monarchy without crowns. Even in the business realm, clothing carries subtle messages of prestige, power, and social intent.

Plants and Indigenous Textiles

Arriving in Samoa in 1845 clad in a thick woolen naval uniform, Commander Charles Wilkes of the U.S. Exploring Expedition found the leaf skirt, or *titi,* of the Samoans nothing less than remarkable. He described the *titi* as

> a short apron and girdle of the leaves of the *ti* (*Cordyline terminalis* [Agavaceae]) tied around the loins and falling down to the thighs. The titi . . . extends all around the body; it has a neat and pretty effect when first put on, but requires renewing often, as the leaves wilt in a few days; this garment is well adapted to the climate, being cool, and the necessity of frequent change insures cleanliness.

Wilkes also found that the Samoans, like other Polynesians, produced bark cloth from *Broussonetia papyrifera* [Moraceae], dyed brownish with the bark of *Bischofia javanica* [Euphorbiaceae], red with the seeds of the neotropical annatto (*Bixa orellana*), yellow with the roots of *Curcuma longa* [Zingiberaceae], or black with tattoo dye from the seeds of *Aleurites moluccana*. Patterns were either painted on by hand or stamped on with carved wooden plates. The soft and flannel-like bark cloth (called *siapo* in Samoa) was worn as a wraparound garment by both men and women on ceremonial occasions.

Bark cloth is still invariably made by women, who find that the process provides an opportunity for pleasant socialization. "In the manufacture of [bark]

Left: Mele Ongeongetau, a Tongan woman, gathers bark from the paper mulberry tree *Broussonetia papyrifera,* which is used throughout Polynesia to manufacture a type of cloth known as "tapa cloth." Right: In Tonga, a woman making tapa cloth wets the bark of the paper mulberry tree and beats it with a mallet on a wooden anvil to spread out the fibers.

cloth, the females of all ranks were employed," wrote William Ellis, an early missionary in the Austral Islands at the beginning of the nineteenth century.

> The queen, and wives of chiefs of the highest rank, strive to excel in some department—in the elegance of the pattern, or the brilliancy of the colour. They are fond of society and work in large parties, in open and temporary houses erected for the purpose. Visiting one of the houses as Eimeo, I saw sixteen or twenty females all employed. The queen sat in their midst, surrounded by several chief women, each with a mallet in her hand, beating the bark that was spread before her. The queen worked as diligently and cheerfully as any present.

The prominence of fiber plants in indigenous societies can be adduced from the number of plant forms they cultivate and from the richness of the ethnotaxonomy (indigenous classification system) they use to classify them. The Maori recognized more than 53 varieties of New Zealand flax, *Phormium tenax* [Agavaceae], known in Maori as *harakeke,* from the leaves that yielded the fibers they wove together to produce a soft, silklike cloth. In March 1770, Sir Joseph Banks, botanist on Captain James Cook's first voyage, wrote with enthusiasm about *Phormium:* "Of all the plants we have seen among these people that which is the most excellent in its kind, and which really excels most if not all that are put to the same uses in other countries, is the plant which serves them instead of hemp and flax."

A list of *Phormium tenax* ethnovarieties reveals not only the tremendous importance that flax played in the Maori culture but also some general principles of ethnotaxonomy. In a paper published in 1973 Brent Berlin, Douglas Breedlove, and Peter Raven argued that all indigenous classification systems have five hierarchically arranged categories. Just as Western scientists categorize plants by kingdom, order, family, genus, and species, Berlin and his colleagues wrote, indigenous peoples generally classify their plants in five categories. The highest postulated category, "unique beginner," corresponds roughly to the Western concept of kingdom. In this as in other categories, the rank may be recognized but not always named. Berlin and his colleagues suggest that "in folk taxonomies it is quite common that the taxon found as a member of the category unique beginner is not denoted linguistically by a single habitual expression. That is, the most inclusive taxon, e.g., *plant* or *animal,* is not named." The next lower category in the Berlin hypothesis, "life form," usually groups plants

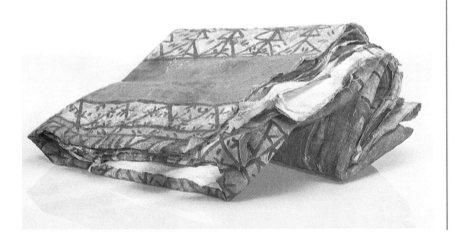

Bolts of bark cloth exceeding 10 meters in length are commonly manufactured in Tonga for ceremonial exchange.

Ethnotaxonomy of Maori Flax
(Phormium tenax)

Generic term: *Harakeke*

Specific terms:

Aohanga	Rātāroa
Atemango	Rauehu
Ateraukawa	Raumoa
Atewheke	Rerehape
Huiroa	Rongotainui
Huki	Ruatapu
Huruhuruwhika	Rukutia
Karhuāmoa	Taeore
Karumanu	Takirikau
Katiraukawa	Tamure
Kauhangaroa	Taneāwai
Kohuinga	Tapoto
Kohunga	Taroa
Kōrītawa	Wharanui
Maomao	Whararahi
Mataroa	Tihore
Motuaruhi	Tīkā
Ngutunui	Tipareouni
Ngutu-parera	Titoonewai
Oue	Tituao
Parekawariki	Toitoi
Parekoritawa	Tukura
Paritaniwha	Tutaemanu
Pehu	Tutaewheke
Pīkōkō	Wini
Potango	

that have the same gross structural features, such as *tree*, *shrub*, and *liana*. The next two categories in the Berlin et al. scheme roughly correspond to genus and species, and the ultimate level includes varieties.

Berlin, Breedlove, and Raven's paper is important not because their conclusions are universally accepted but because their suggestion catalyzed a major discussion in both ethnobotany and cultural anthropology and inspired researchers to take a closer look at indigenous naming systems. It appears that most ethnotaxonomic systems are similar to the modern scientific system in their use of binomials to name plants. Just as in the modern botanical systems, a binomial in an ethnotaxonomic system consists of a generic-level term and a specific modifier. As we have seen, the generic Maori term for New Zealand flax is *harakeke*. Adding the specific modifier *huiroa*, which translates as "long meeting," to the generic term produces *Harakeke huiroa* ("long meeting flax"), a name applied to an especially fine cultivar suitable for making exquisite garments for ceremonial display.

Ethnobotanists who studied ethnotaxonomic systems soon discovered that the way a thing is categorized varies from culture to culture. As we mentioned earlier, the ethnobotanist Nina Etkin found that the Hausa of West Africa make no clear distinction between medicine and food, an insight that has proved to have crucial significance for the study of medicinal plants. An even more surprising discovery was made by Christin Kocher Schmid of the Ethnologisches Seminar in Basel, Switzerland. Working in the Madang province of Papua New Guinea, she found that the Nokop people conflate categories of plants used for objects that English speakers consider quite distinct—"string" and "skirt," for instance. Plants used to weave string and plants used to make skirts are all considered to belong to a single category, called either *male silep* or *kalak silep* (*silep* = family; *male* = string material; *kilak* = skirt material). The reason is quite simple: the Nokop people wear skirts woven from string.

Fishing Implements: The Vanishing Art of *'Enu* Weaving

Maori flax is abundant throughout New Zealand, as are the plants used in New Guinea to weave string skirts. But what happens when a plant needed for a community use becomes scarce? This question is becoming of ever greater concern, as clear-cutting and hunting by outsiders are destroying natural resources at rates never before seen. Of course, indigenous peoples have always had to adapt to scarcity. In the case of the *vesi* trees in Fiji, the scarcity of large trees elsewhere in the region gave tiny Kabara island a political saliency out of proportion to its other natural resources as shipwrights from other islands took up

The *'ie'ie* vine *Freycinetia reineckei* relies on the Samoan flying fox *Pteropus sanoensis* to move the pollen from the male plants to the female plants. Drastic declines in the flying fox population due to hunting, logging, and cyclones have reduced pollination of the vine, which reduces the potential supply of *'ie'ie* roots needed to weave *'enu* fish traps.

residency there. What is new today is the suddenness and rapidity of change. In the modern world, the scarcity of a crucial plant can lead not so much to adaptation but to the loss of a vital part of the culture. Such threatens to be the case with fish traps (called *'enu*) in Samoa. In the Manu'a islands of Samoa, the vines used to weave fish traps are disappearing. With the disappearance of the vine, a rich part of Samoan culture may vanish.

Only three living Samoans know the art of weaving *'enu*. All live in Manu'a, a group of three islands approximately 100 miles east of the capital city of Pago Pago. One of the weavers, Leatioti Tauluava'a, lives on Olosega, a vegetation-clad volcanic island with 2000-foot cliffs that seem to leap straight up out of the Pacific Ocean. A visit to Leatioti's village reveals *'enu* that are generations old hanging in the huts: very few new ones have been constructed. "It is not possible to weave them anymore," Leatioti explains. "The *'ie'ie* vine no longer flourishes on the island. It seems to be disappearing for some reason. Perhaps it exists on top of the mountain, but now I am too old to climb up there."

Interested in learning about Leatioti's weaving techniques, Paul Cox and his students climbed the central mountain of Olosega, using a small path cut in the cliffs, in search of *'ie'ie* vines. But after an entire day of hunting, they found only a few vines. Even if Leatioti were young enough to climb the mountain, he would still find it impossible to weave *'enu*.

The Samoan flying fox *Pteropus samoensis* [Ptoropididae] is the major pollinator of the *'ie'ie* vine (*Freycinetia reineckei*) used to weave fish traps. The near extinction of the flying fox has led to a disappearance of the vines and a loss not only of the weaving techniques but of the legends, songs, and rituals that accompany the harvest of the migratory *i'a sina* fish.

Where are the *'ie'ie* vines? These vines, *Freycinetia reineckei* [Pandanaceae], grow in the primary forest of Samoa. Several recent hurricanes have severely damaged the Olosega forest, but the real reason for the loss of the *'ie'ie* vines is more pernicious: the vines have lost their pollinator, the Samoan flying fox, *Pteropus samoensis,* and so are unable to produce seed. Thus few new vines are growing to replace old and damaged ones. Once flying foxes were a common sight on Olosega, soaring high above the forest on the afternoon thermal winds on wings spanning 4 feet or more. The large, sweet bracts of the *'ie'ie* flowers were their favorite food. But during the late 1980s commercial hunters arrived from Guam with semi-automatic weapons. Within three years, 18,000 dead flying foxes were shipped from Samoa to Guam, where they were sold as a luxury food item. The U.S. government refused to ban the commerce in them in Guam, so by the time the Samoan government prohibited their export, the flying fox population was teetering on the brink of extinction. Two successive hurricanes reduced their population to less than 5 percent of earlier levels. Although the few remaining flying foxes may still reproduce their species, it may not be possible for them to fulfill their roles as effective pollinators of the *'ie'ie* vine before the vines have disappeared.

Without the long slender roots of the *'ie'ie* vine, it is impossible to weave *'enu*; and without the *'enu*, it is impossible to catch the sardine-like *i'a sina* fish that migrate at certain times of the year. Even the feasts, songs, and legends surrounding the *i'a sina* migration will disappear without functioning *'enu*.

In an attempt to preserve at least the knowledge of how to weave *'enu*, Cox and his students traveled to Savaii island, 300 kilometers distant, and together with Falealupo villagers (none of whom know how to weave *'enu*) gathered precious *'ie'ie* roots. They returned to Olosega and presented Leatioti with their bounty: 20 kilograms of *'ie'ie* roots. Leatioti was delighted and began weaving an *'enu* immediately. He explained that the *'ie'ie* is the only material used for the ribs of *'enu* because only its roots can withstand repeated immersion in seawater. Typically, he explained, the roots are collected from the vines that grow in the primary rain forest and taken to the village, where they are scraped free of any obtrusions. The roots are immersed in seawater for several days, then coiled, dried, and stored until the time comes to weave an *'enu.*

The *'enu* is woven in two parts: the basket, or trap, and the funnel that forms the entrance. The *'enu* is lashed with sennit made of coconut fiber for strength. The trap is placed horizontally with a little crabmeat inside it for bait. When the *i'a sina* are migrating, a single trap can yield several kilos of fish.

The ethnobotanists positioned still and video cameras and audio recorders to document Leatioti's weaving process, and the attention lavished on him soon attracted the interest of the village youth. The ethnobotanists were pleased, for they hoped that one of the young men would become Leatioti's apprentice, to learn his technique and pass it on to future generations.

The loss of *'ie'ie* vines from Olosega provides a confirmation of the profound influence that plants have on human cultures, since the survival of a small but significant part of Samoan culture hangs on the survival of the vine, including the *'enu* weaving and fishing techniques; the songs, poetry, and leg-

Only two living Manu'ans remember how to weave the *'enu* fish traps from *'ie'ie* roots. The fish traps, when properly used, can capture kilograms of the small fish in only a few hours.

ends associated with the *i'a sina* harvest; and the socialization processes inherent in *'enu* weaving and communal consumption of the catch. In this case, as in many others, biological and cultural conservation are interrelated, for it is the slaughter of the plant's primary pollinator, the Samoan flying fox, that has threatened this important part of Samoan culture.

The loss of *'ie'ie* vines from Olosega illustrates, too, the economic effects of plant scarcity. Although the *'ie'ie* vine itself is not consumed by the islanders, its loss could result in a reduction of the number of people the island can support, since the islanders can no longer efficiently harvest the *i'a sina* fish.

Some island plants have had profound effects on island cultures not because they are scarce but because they are of sufficient local abundance to be exported. Such is the case with *Myristica fragrans* [Myristicaceae], a tree originally found only on the six small islands of Banda (total area 40 square miles [104 square km.]) in what is now the Maluku province of Indonesia. If there was ever a plant that became the dream of entrepreneurs and a siren for navigators, it is the fruit of this tree from which were built the fortunes of Genoa, Venice, Lisbon, Madrid, and Amsterdam. The story of *M. fragrans* is far more than the story of a desirable plant, however, for it illustrates how indigenous cultures can be deeply affected, and even nearly exterminated, if their plant resources assume an importance in distant lands that is disproportionate to their local value.

While languishing in a Genoan jail in the years 1298 to 1299, Marco Polo ignited the first spark of European interest in this plant. Together with the romance writer Rustichello of Paris, he wrote the *Travels,* a florid chronicle of his 20-year journeys in the Mongol empire and beyond. As an adventure story, the *Travels* was at the time and remains today a very good read. But what seized the imagination of Marco Polo's countrymen was his account of an island where a plant grew whose fruit approached the value of gold: nutmeg.

Plants as an Impetus for European Exploration

Surrounded by a mirrorlike sea, Banda Neira seems like one of the most tranquil spots in the world. With rainfall of 221 to 367 centimeters per year and a rich volcanic soil, the island is a veritable paradise. Yet on the lower slopes of its volcano, Gunug Api, is one of the most formidable stone fortresses in the Southern Hemisphere. The construction of this citadel, with turrets and cannon aimed at the bay below, is far beyond the abilities of even the modern Bandans, who still sail sea craft resembling Fijian *camakau* among their islands. Closer inspection of the fort reveals plaques inscribed not in the Bandan or Malay language, but in the Dutch script of the seventeenth century.

Across the bay are orchards of *M. fragrans*. Most of the bushy evergreen trees stand 9 to 12 meters high. Their fruit, locally called *pala gula,* is yellow, smooth, and fleshy, something like an apricot. The islanders eat the dried fruit as a confection. If the sole production of *M. fragrans* were *pala gula,* Banda's history would have been far more tranquil than it has been. Certainly there would be no stone fort and no cannons. But at maturity, the fruit splits open to reveal a secret at once mysterious and compelling. There, exposed to view, is a dark-brown nut covered by a clawlike crimson aril, or fleshy seed covering. When the aril is dried, it becomes the spice we call mace. When the nut is cracked, it yields a large kernel which we call nutmeg.

For thousands of years the nutmeg tree could be found only in Banda, "close enough to smell the sea," as the Bandans are fond of saying. Yet despite an ancient prophecy that warned of fair-skinned people who would overrun Banda and destroy its people, few of the *orang kaya,* or village elders, associated commerce in nutmegs with an impending apocalypse. For centuries the islanders traded nutmegs with nearby archipelagos. Indeed, the nutmeg seems ideally containerized for transport, protected by a hard, waterproof outer shell and filled with anti-oxidant compounds. The ground spice itself can be stored for years without loss of flavor. The transportability and very limited natural range of nutmeg made Banda the center of a commerce that structured the

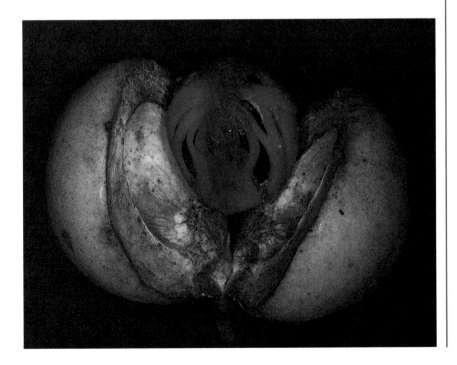

The fruit of the nutmeg tree *Myristica fragrans* yields two valuable spices. The inner black kernel produces nutmeg, while the red clawlike aril surrounding it produces the spice called mace. The Banda islanders eat the outer husk as a confection.

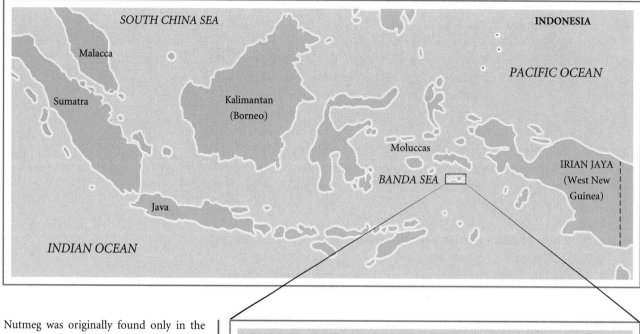

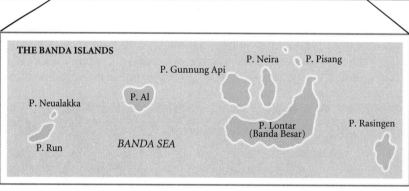

Nutmeg was originally found only in the Banda Islands, located in Indonesia to the west of New Guinea. The possibility of obtaining fame and fortune by capturing these spice islands spurred European navigation and voyaging.

geopolitical trajectory of the Western world in a way even the Persian Gulf only remotely approaches in recent times.

As early as the fifth century, Banda's nutmegs reached India. By the ninth century this trade had become more formalized as Indian traders based in Java established a monopoly. Arabian sailors started shipping nutmegs to Charx, on the Persian Gulf, and to Egypt through the Red Sea from India. The Arabians were very secretive about the source of the spice and soon built a flourishing spice monopoly that reached as far as Constantinople. Many a curious adventurer returned from Arabia disappointed by his failure to find nutmeg trees growing in the desert. Europeans' contact with Arabia was limited during the early Middle Ages, but the Crusades reopened the trade routes between the East and the West, and nutmeg and cloves (*Syzygium aromaticum* [Myrtaceae]) became extraordinarily popular. Merchants from Venice and Genoa resident in Palestine traded clothing and iron to Crusaders for spices. The resultant flow of spices northward became so copious that in 1191 the Romans fumigated the streets with nutmeg for the coronation of Emperor Henry VI.

Both Venice and Genoa became enormously wealthy as nutmeg and mace flowed north from these cities throughout Europe, even reaching the British Isles. In 1284 a pound of mace sold for the value of three sheep in England, and in the fourteenth century Geoffrey Chaucer recorded the use of nutmeg to flavor English ale. Still the origins of nutmeg and mace were shrouded in mystery, until Marco Polo, from his prison in Genoa, described whole orchards of nutmeg trees growing on distant islands. His account spurred a search for these "spice islands."

They were not easily found. With its defeat of Genoa in 1380, Venice monopolized the Middle Eastern spice trade with Europe. The value of nutmeg spiraled upward when Constantinople fell to the Turks in 1453. The overland route from the Middle East was further complicated by a tax imposed on nutmegs by the sultan of Egypt. A few adventurous souls began to glimpse the potential reward for anyone who discovered a direct sea route to the Spice Islands.

The seeds of doom for the Venetian spice monopoly sprouted with the establishment of a school for navigators on the Iberian Peninsula. On July 8, 1497, the Portuguese explorer Vasco da Gama left Lisbon with four ships, determined to pioneer a direct sea route to the spice markets of India. He reached Calicut, on the Malaban coast of India, in 1498 and returned to Lisbon in 1499 with a shipload of spices. Another Portuguese expedition commanded by Pedro Alvares Cabral left Lisbon in 1502 and returned the next year with all the ships laden with spices. The resultant price fluctuations in nutmeg, mace, and cloves shook the Venetian spice empire to its core. In 1503 the Portuguese king sent Alfonso de Albuquerque to India, where he built a fort at Cochin; by 1506 the European spice trade had shifted to Lisbon and Venice's monopoly was shattered. Determined to eliminate all competitors, Albuquerque destroyed the Arabian trade at its source in 1510 and the next year captured Malacca, on the Malay Peninsula. His conquest secure, Albuquerque immediately sent three vessels in search of the Spice Islands.

The expedition sailed southeast and arrived at Banda in 1512. They loaded their ships with nutmegs, mace, and cloves purchased from the islanders at prices that were guaranteed to produce profits that would provoke the envy of

An islander gathers fruits from a nutmeg tree in Banda. The seizure of the islands by European powers hoping to control the nutmeg trade nearly led to the annihilation of the islands' inhabitants.

modern cocaine merchants. The enormous profitability of the spice trade moved one of the crewmen to approach the Spanish monarch on his return to Europe. His name was Ferdinand Magellan.

As one of the few living Europeans to have visited Banda, Magellan enjoyed a high degree of credibility in the Spanish court, but he faced a serious obstacle: in 1494 Pope Alexander VI had ceded all "heathen" lands east of Europe to Portugal. Undaunted, Magellan made a suggestion that the Spanish court had already viewed with favor some years earlier when a young Genoan navigator had proposed it: like Columbus, he would reach the East by sailing west.

Magellan departed Sanlúcar de Barrameda on September 20, 1519, with five ships and a Spanish crew of 270. After wintering in Patagonia, they pressed on. One ship ran aground and another deserted before the expedition rounded the southern tip of South America and passed through the straits that later would bear Magellan's name. Reaching the Philippines in 1521, Magellan and 40 of his men were killed in a battle with the local people.

Too few sailors were left alive to man all three vessels, so one ship was abandoned. The remaining two ships, the *Trinidad* and the *Victoria,* made their way to Banda under the command of Juan Sebastián del Cano. At Banda they took on such a large cargo of nutmegs that the *Trinidad* ran aground and its crew was captured by the Portuguese. In May 1522 the *Victoria* rounded the Cape of Good Hope, at the southern tip of Africa, and sailed with its starving crew to the Cape Verde Islands, then firmly in Portuguese hands.

The crew of the Victoria gambled that they could convince the Portuguese garrison that they were returning from Spanish colonies in South America. The Portuguese took pity on them until one foolish crew member paid for a purchase with the most valuable commodity he possessed: a handful of cloves. Realizing that nutmegs, mace, and cloves could not be found in South America, the Portuguese gave pursuit. The Spaniards still on board immediately put to sea, and they limped into Spain on September 8, 1522. The original crew of 270 was now down to 17 men. Yet the expedition was far from a failure: the spices in the *Victoria*'s hold more than repaid the king's investment. He awarded Cano a coat of arms emblazoned with a nutmeg.

Cano's circumnavigation of the globe under the Spanish flag did little to reduce Portugal's hold on the spice trade. In 1580, however, Philip II of Spain claimed the Portuguese throne by virtue of his brief marriage to a Portuguese princess 37 years earlier, becoming King Philip I of Portugal. Now Spain and Portugal were united under one crown.

Having previously inherited sovereignty over the Netherlands, Philip I set the political stage for a profitable relationship between the Iberian Peninsula and his Dutch possession: Spain and Portugal would import nutmegs from Banda and the Netherlands would market them throughout northern Europe. Again, however, the lucrative monopoly on the spices would be short-lived.

In 1595 a group of Amsterdam spice merchants decided to eliminate their Portuguese intermediaries by sending an expedition directly to Banda. The Dutch expedition returned in 1598 with 45 tons of nutmegs and 30 bales of mace, an amount sufficient to guarantee all of the original investors fabulous wealth for life. The first Dutch expedition did not arouse the concern of the *orang kaya,* or village elders, in Banda, but they did not view the arrival of Vice Admiral Jacob van Heemskerk in 1599 as propitious because it coincided with ominous activity of the volcano Gunug Api. The *orang kaya* wondered if the Dutch were the prophesied destroyers of their people.

Van Heemskerk announced that the Dutch were enemies of the Portuguese and wished to supplant them as Banda's trading partners. He presented gifts to the *orang kaya* and began to barter mirrors, cloth, gunpowder, and knives for nutmegs and mace. The prices the Bandans demanded were high, but still nutmegs that cost the Dutch the equivalent of $1 in Banda could be resold for $30,000 in Amsterdam. The spice trade promised to usher the Netherlands into a new era of economic prosperity. However, news soon spread of an attempt by "English gentlemen" to sponsor a similar series of expeditions to the Spice Islands. Readers of John Gerard's *Herball* in 1597 learned that "the nutmeg tree groweth in the Indies, in an island especially called Banda." Clearly more nations than just Portugal and Spain had their eyes on Banda.

Recalling the vanquishment of Venice's spice trade by the Portuguese, the Amsterdam merchants determined that the Netherlands alone would control Banda. They sought a charter for a new company equipped with both diplomatic powers and military might. In 1602 the Vereenigde Oost-Indische Compagnie (Dutch East Indies Company, or V.O.C.) was created, with authority to wage war against Spain, Portugal, and other foreign countries in the Spice Islands.

On May 23, 1602, Admiral Hermanszoon of the V.O.C. coerced the *orang kaya* of Banda to cede an irrevocable monopoly on nutmegs and mace to the Netherlands. Soon quasi-military forces under the V.O.C.'s control drove the Portuguese from Banda and Goa and blockaded Malacca and Java. A Dutch expedition in 1605 was ordered to permanently expel the English from all of the Molucca islands.

On April 25 the *orang kaya* watched the ancient prophecy unfold as 750 Dutch soldiers landed and began to construct a stone fort. This show of military might convinced the elders that the Dutch were indeed the prophesied fair-

The Portuguese began construction of this castle overlooking Banda harbor, but it was later seized by the Dutch to maintain the monopoly on nutmegs.

skinned conquerors. They decided to take action. On May 22, 1609, the Bandanese under the direction of the *orang kaya* ambushed the Dutch admiral and 20 of his staff. When it was all over, they hoisted the admiral's head on a battle lance for public display.

Retribution was slow but harsh, arriving in the person of Jan Pieterszoon Coen, governor general of the V.O.C. He sailed to Banda in 1621 with 16 warships, 36 barges, 1905 soldiers, and 100 Japanese mercenaries, among them some trained executioners. The *orang kaya* sued for peace, but on April 21, 1621, a genocidal war commenced. The Dutch burned all the villages and slaughtered the inhabitants they considered unsuitable as slaves. The rest they shipped to Java. One consignment of slaves consisted of 287 men, 356 women, and 240 children. Coen's brutal show of force horrified even his officers. "The forty-four prisoners were brought within the castle," reported Lieutenant Nicolas van Waert.

> The condemned victims being brought within the enclosure, six Japanese soldiers were also ordered inside, and with their sharp swords they beheaded and quartered the eight chief *orang kaya* and then beheaded and quartered the thirty-six others. This execution was awful to see. The *orang kaya* died silently, except that one of them, speaking in the Dutch tongue, said "Sirs, have you then no mercy?" . . . The heads and quarters of those who had been executed were impaled upon bamboos and so displayed.

After nearly exterminating the native population (of the original 15,000 inhabitants, only 1000 remained), Coen returned to Batavia and announced that the V.O.C. would accept applications from Dutch citizens for grants of land in Banda if they would agree to live there permanently. The V.O.C. would provide slaves to work the nutmeg orchards, supply food at cost, and purchase all nutmegs at a fixed price. In an etymological coincidence, the people who received these perquisites were called "perkiners."

The perks of the spice trade were indeed tremendous. The perkiners began to live in ostentatious splendor. As their gambling debts mounted, however, they began to adulterate their shipments of nutmegs with the seeds of other species of *Myristica* that grew on many islands other than Banda. The presence of these so-called long nutmegs in shipments of the true spice so infuriated the V.O.C. that the company dispatched expeditions to extirpate all *Myristica* species from all the surrounding islands. Such efforts became the botanical equivalent of Sisyphean labor, because the bright-red aril of all species is highly attractive to birds, and they routinely fly among islands with the fruit in their beaks. The perkiners' subterfuge had serious consequences for the market: it trained consumers to accept an inferior product. "Long nutmegs" could be obtained on many tropical islands, so it was not long before merchants from other countries were trading in them.

The Dutch monopoly lasted for nearly a century, but adulteration of the spice and corruption within the V.O.C. made the Dutch vulnerable. Two intrigues, one by the French and a second by the English, broke the Dutch monopoly. In 1770 the French botanist Pierre Poivre of Mauritius smuggled nutmeg and clove trees out of the Moluccas to Madagascar and Zanzibar, where they were successfully cultivated. The British approach was far less discrete: on February 7, 1796, the Dutch garrison surrendered to an English fleet that invaded Banda under the pretext of Dutch–English rivalry in the Napoleonic Wars. The English not only seized spices but transported nutmeg trees to British soil in Sri Lanka, Malaysia, and eventually Singapore. Amsterdam spice merchants burned nutmegs in the streets in protest, but the Dutch spice monopoly was irretrievably lost.

The nutmeg production in Banda has long since been eclipsed by production in Grenada, in the Caribbean. Its importance as a spice and a preservative continue to guarantee an ongoing demand. But a novel use of it emerged in the 1960s: nutmeg was claimed to be a hallucinogen.

In prisons throughout the United States, word spread among bored inmates that it was possible to get high by ingesting large doses of nutmeg. The prisoners' experiments were not entirely without precedent: anecdotal accounts of the hallucinogenic effects of nutmeg can be traced as far back as 1576, when Lobelius reported "a pregnant English lady, who, having eaten 10 or 12 nutmegs, became deliriously inebriated." Rumphius in 1741 recounted a tale of two soldiers who slept underneath a nutmeg tree and woke up feeling drunk. The accounts of prisoners and hippies in the 1960s are inconclusive: some reported auditory and visual hallucinations from nutmeg use, but others found no effect.

The determination of nutmeg's putative hallucinogenic properties is of some practical value: since large doses of nutmeg are potentially lethal, experimentation by curious young people entails serious hazard. Reasoning that any psychoactive properties of nutmeg were unlikely to have escaped the Bandans' notice, Dutch-speaking ethnobotany student Carl Van Gils traveled to Banda and the surrounding islands in the Maluku province.

Today the Bandans grow nutmeg trees both near their homes and in old plantations. They pick up the fruit from the ground or use a *gaigai*, a long stick to which they have attached a hook and a basket, to pluck it from the trees. Typically they wait until the fruit has split open before they pick it.

As we mentioned earlier, the fruit called *pala gula* is dried and eaten as a confection. The people of Maluku use nutmeg oil to treat flu by rubbing it all over the body to produce a warm, strengthening feeling. They combine grated nutmeg with eucalyptus oil and strap it on the abdomen to treat diarrhea. Nutmeg also continues to have religious significance for Bandans: they tie a nutmeg around the neck of a seriously sick infant for whom no other cure can be found, and ask God to heal the child.

Farther away, in Java, Van Gils found that mace is combined with other herbal ingredients and sweetened to produce *obat penenang,* or "calming medicine." The drink is claimed to alleviate insomnia, stress, and nervousness. Throughout Indonesia, grated nutmeg in warm milk is used to help infants and toddlers sleep. But neither in the Maluku province nor in Java is nutmeg used as a hallucinogen. The most distinguished ethnobotanist in Indonesia, A.J.G.H. Kostermans, told us that nutmeg is not used as a narcotic in Indonesia. Richard Evans Schultes and Albert Hoffman, the inventor of LSD, tested the most active component of nutmeg, myristicin, for psychotropic effects and concluded that the only psychoactive effects of nutmeg are manifestations of general toxicity. Since the levels of toxicity are "extremely variable," nutmeg cannot be considered a hallucinogen. It is a pseudo-hallucinogen, one of a class of substances whose psychoactive effects, if any, is obtained only at near-toxic doses. Where such plants are concerned, then, the line between hallucinations and death is very thin.

Economic traffic in plants such as nutmeg has indeed had a profound impact on human cultures. But some of the deepest impacts are those made by plants used for religious and spiritual purposes, impacts that can only rarely be measured in economic terms. Such plants are used to structure and express the relationship between this life and the next, and many cultures have expressed their conception of humankind's place in the universe with plants. Consider, for example, the Maori *tupapaku* and the Maori *wakahuia*.

Plants and Human Cosmologies

The Maori *waka tupapaku* (*waka* = boat; *tupapaku* = corpse), carved from the end of a canoe, is much smaller than the Fijian *camakau,* because it was designed to transport a single passenger to the world of spirits. Just as the great

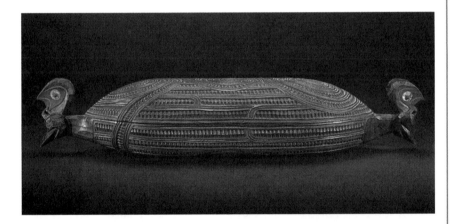

The Maori *waka huia* is intricately carved with spirals representing the unfolding fiddle head of a tree fern, a symbol of life and renewal after death. But the notches in the spirals could alternatively represent not microcosm but macrocosm: stars embedded in the visible galaxy. Although the carving of the exterior (top) represents considerable craftsmanship, the interior of the box (bottom) is filled with random slashes, showing that the human condition within the universe is indeterminate, replete with both risk and possibility.

Maori canoes brought their ancestors from the East, so *waka tupapaku* transferred the dead to a better world. Although unseen, portals to the next world
could be fashioned from plants. Such portals became *tapu* (taboo): holy, sacred,
and forbidden to all but chiefs and priests.

The Maori *waka huia*, a box carved from miro wood (*Podocarpus ferruginea*
[Podocarpaceae]), is scarcely half a meter long, but it was designed for voyages
not to distant islands but to the stars and beyond. *Huia* is the Maori name for a
now-extinct bird. The *waka huia* was used to hold the feathers used to orna-
ment the hair of chiefs. The intricate carvings on the container, however, are far
more than decorative: they have spiritual and cosmological significance. On the
top and bottom of the box are carved double spirals that look like the curling
young fronds of a New Zealand tree fern, *Cyathea dealbata* [Cyatheaceae], and
so represent new life and renewal after death:

> Ka mate he tete,
> ka tupu he tete.

> As one fern frond dies,
> another fern frond grows.

Death and the renewal of life as symbolized by a fern frond was a potent image for the Maori people, but even deeper layers of meaning can be found in
the carvings on a *waka huia*. Tiny notches carefully cut along each spiral afford
an alternative interpretation for a people who navigated by the stars from their
ancestral home of Havaiki: they can represent stars in the large-scale structure
of the universe. Carved heads of ancient deities appear at either end, with
mother-of-pearl eyes that iridesce in the slightest light. The message is clear:
God is at the beginning and the end of the universe, alpha and omega, and in
between the universe is, in Freeman Dyson's phrase, "infinite in all directions."

The late Maori ethnologist Te Rangi Hiroa explained that in the Maori cos-
mology, Ranginui, the sky father, loved Papatuanuku, the earth mother, who lay
naked on her back facing him. To cover her nakedness, Rangunui placed plants
on her head, her body, and her armpits. He caused great forests to grow on
Papatuanuku. Maoris regarded the rain as tears shed by the sky for love
of the earth. Early in the morning one sometimes sees the rising mists
that are Papatuanuku's embrace of love for the sky.

When the *waka huia*, this indigenous model of the universe, is opened, the
Maori predilection for the future is manifest: we behold a primitive and almost
random pattern of broad slashes. The carver's intent is clear—within the large-
scale structure of the universe, the human condition is indeterminate. As befit-

ted the first people ever to set foot in Aotearoa, that biological paradise we call New Zealand, the Maori believed the future to be inchoate, full of both risk and possibility.

Why should a small carved box hold such fascination for an ethnobotanist? Because a single artifact of material culture can reveal deep cultural significance. Rare indeed is the manufactured object that acquires such significance in Western culture. Even a Fabergé egg, delightful as an intricate artifice, fails to inspire the religious awe of a *waka huia*. And a crucifix, potent religious symbol though it is, tells us little about the possessor's culture. Imbued with both the sophisticated craftsmanship of a Fabergé egg and the religious potency of a crucifix, a *waka huia* contains something more: *mana,* or spiritual power. Even today some of our Maori friends would not enter our house if they knew that a *waka huia* was present.

Because a *waka huia* was *tapu,* it could be used only by a chief. Nearly everything relating to a chief was *tapu:* his house, his clothes, his possessions, his food. Those who defiled a *waka huia,* whether by touching it without the requisite authorization, gazing upon it, or even speaking disrespectfully of it, became subject to serious illness or injury unless they immediately sought ritual purification. For this reason, a person who wished to carve a *waka huia* had first to be purified by a master carver. The master took the person naked into a river at the setting of the sun, sprinkled water on his head, and spoke the sacred words of an ancient chant.

It is perhaps this religious realm of indigenous plant use that is the most difficult for Westerners to appreciate or even understand. Immersed in what Yale professor Stephen L. Carter calls "the culture of disbelief," many modern Americans and most modern Europeans have only a superficial and distant relationship to religion. And yet to indigenous peoples, the spiritual realm is not only real, but pervasive, and structures not only their views of the universe, but much of their behavior in this world. Here again plants play a crucial role, for they are often the key that opens the door to the other world.

In this painting, entitled *The Induction of the Ayahuasca in the Brain*, Peruvian shaman and artist Pablo Amaringo illustrates a vision he experienced after imbibing a beverage made from the vine ayahuasca, in which he sees the Ayahuasca Mother transformed into a Shipibo Indian woman with many eyes, eyes that are good (with joy and peace) as well as bad (with seduction and lack of generosity).

Entering the Other World

Spice plants proved to be the lure that drew Europeans to explore and colonize the known world. Many indigenous peoples also use a broad range of plants to open channels to another world, a world perceived to be inhabited by spirits: by the spirits of those who have died, by benevolent deities, and by malevolent demons. The influence of these spiritual entities is not confined to that other world; indeed, they can have powerful influence on the affairs of this world. The person who wins their favor may reap health, wealth, and political power.

But not all contact with the spirit world is beneficial to everyone, nor is the use of plants to open the channel between worlds universally advantageous. Ritual use of the "ordeal bean of Calabar," *Physostigma venenosum* [Fabaceae], for example, had a terrible effect: nearly all of the people who ate the bean died. In Duke Town, Calabar, West Africa, so many people died after eating

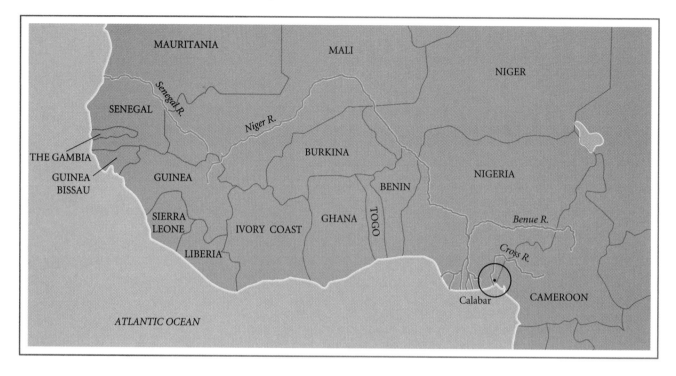

Calabar, a region on the west coast of Africa, is traditional home to the Egbo, a secret society of the Efik people that used the Calabar bean *Physostigma venenosum* to conduct trials by ordeal.

Calabar beans in the mid–nineteenth century that the British consul issued an emergency ordinance: "Any person administering the [Calabar] bean, whether the person taking it dies or not, shall be considered guilty of murder and suffer death."

Why would anyone administer such a potent poison to a human being? To answer such a question, it is necessary to explore the interface between plants and ritual, the profane and the sacred. Pharmacologists can be content to document the toxic properties of the Calabar bean; chemists can elucidate the molecular structures of the toxins; systematic botanists can clarify the taxonomic affinities of *Physostigma venenosum*; but ethnobotanists must attempt to understand the plant's role in the indigenous culture. And to unravel this mystery, they have to investigate indigenous views of both the angelic and the demonic, and understand through indigenous eyes the spirit occupants of the chasm between this life and the next.

The plants used in rituals and religious rites often provide keys to indigenous cosmologies. Just as Westerners incorporate flowers in their funeral observances, indigenous peoples use plants to symbolize the link between this world and the next. But many indigenous peoples believe that some plants function not merely as symbols of the next life but as conduits of spiritual power or even *as* gods. One may gain access to the world of spirits by participating in rituals involving plants, such as those of the Dionysian cult of ancient Greece, which used fennel (*Foeniculum vulgare* [Apiaceae]); by ingesting a psychoactive plant, as in the peyote ceremonies of the Native American Church; or by membership in a secret society that uses plant poisons to ascertain if a person in their midst is a witch. Such faith in the spiritual power of plants cannot be dismissed as irrational, for a potent pharmacological reality may underlie it, as in the case of the use of the Calabar bean among the Efik, a tribal group in Nigeria.

For generations Efik nobles traded in slaves and palm oil, and their households grew rich and powerful. Material prosperity was not the sole basis of the nobles' power, however, for they all belonged to the Egbo, a secret quasi-religious society replete with special costumes and ranks. The Egbo made and enforced laws, adjudicated disputes, and in effect formed the government of the Efik. Violation of the Egbo's edicts could lead to a fine, corporal punishment, or death. So feared were the agents who carried out the Egbo's orders that on the society's weekly meeting day, the uninitiated hid indoors to avoid encountering them. A serious offense against the Egbo could result in trial by *esere,* the ordeal bean of Calabar.

The drastic effects of the Calabar bean were quickly noted by visiting Europeans. "The king and chief inhabitants ordinarily constitute a court of justice, in which all country disputes are adjusted," wrote William Daniell, a British medical officer stationed in Calabar in 1846.

> If found guilty, they are usually forced to swallow a deadly poison, made from the seeds of an aquatic leguminous plant. . . . The condemned person, after swallowing a certain portion of the liquid, is ordered to walk about until its effects become palpable. If, however, after the lapse of a definite period, the accused should be so fortunate as to throw the poison off the stomach, he is considered as innocent and allowed to depart unmolested.

Survival of trial by Calabar bean was therefore regarded by the Egbo and Efik society in general as constituting prima facie evidence of innocence, particularly against charges of witchcraft. "The Efik believe that the *esere* or Calabar bean possesses the power to reveal and destroy witchcraft," wrote the anthropologist Donald Simmons in the 1950s. Trial by Calabar bean calls to mind the trials by ordeal suffered by persons accused of witchcraft in medieval Europe. A suspected witch might be bound and thrown into

The Calabar bean, known as *esere* by the Efik people of Nigeria, was used in trials by ordeal to either substantiate or refute allegations of witchcraft.

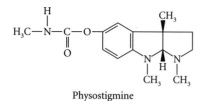

Physostigmine

Physostigmine, an alkaloid isolated from the ordeal bean of Calabar, is an important drug used in the treatment of glaucoma.

deep water. If she floated, she was indeed a witch; if she drowned, she was innocent. Such spectacles might evidence ignorance, superstition, and spiritual blindness but in an amazing turn of events, the ordeal bean of Calabar has contributed new vision to Western medicine—the Calabar bean is the source of physostigmine, a potent drug used to prevent blindness in glaucoma patients.

The Swedish ethnopharmacologist Bo Holmstedt has carefully traced the history of the development of physostigmine from the ordeal bean of Calabar. He reported that Robert Christison, a physician at the University of Edinburgh, dosed himself with a quarter of a seed and experienced giddiness, torpidity, and such loss of volition that his family summoned a doctor. The physician, Christison later reported, "found the pulse and action of the heart very feeble, frequent, and most irregular, the countenance very pale, the prostration great, the mental faculties unimpaired, unless perhaps it might be that I felt no alarm where my friends saw some reason for it."

His countryman Thomas Fraser, an ardent student of African blowgun poisons, followed up Christison's work. He discovered that application of the Calabar bean to the eyeball caused the pupil to contract sharply:

> In about thirty minutes after the application the pupil becomes a mere speck, but still retains a certain degree of mobility. It continues in this state for twelve or fourteen hours, but greater or less degree of contraction of the pupil may persist for five or six days.

Fraser suggested that his friend Douglas Robertson, an ophthalmic surgeon, investigate the effect of the Calabar bean on the eye. Robertson's experiments on his own eyes demonstrated "that the local application of the Calabar bean to the eye induces, first, a condition of shortsightedness . . . and second, it occasions contraction of the pupil."

In 1864 German chemists Julius Jobst and Oswald Hesse isolated the alkaloid physostigmine from the Calabar bean. After the compound was found to reduce pressure in the eye dramatically, it became one of the treatments of choice for glaucoma.

Despite the ultimate utility of the Calabar bean in ophthalmology, its story is replete with human suffering and misery. But plants that cause death are not the only doors to the other world. Indeed, most plants used by indigenous peoples as conduits to the spirit world are used for beneficial purposes. Over 7000 kilometers to the west of Calabar, in South America, another plant imbued with spiritual power was used not to destroy political enemies but to heal the sick by imparting spiritual knowledge for difficult diagnoses.

Ebena Snuff in South America

The Waiká shaman puts a small amount of *ebena* powder into a reedlike snuff tube and inserts one end of it in his nostril. His assistant blows a strong blast on the other end of the 1-meter long tube, propelling the bioactive powder into the shaman's nasopharyngeal airway, where it is absorbed into the bloodstream across the mucous membranes of the respiratory tract. Within 60 seconds the powerful alkaloids in the powder are distributed throughout the circulatory system by way of the bloodstream. At once the shaman begins to scratch the top of his head with a circular motion and saliva pours uncontrollably from his mouth. Within a few minutes the shaman's spirit leaves his body and enters the other world, a world controlled by spirits who act as friend or foe. Today the shaman has been called to help a young boy with a high fever and other symptoms thought to be caused by evil spirits who have taken control of the boy's soul and body. In his quest for a cure, the shaman enlists the help of apparitions that have befriended him, the spirits of the monkey and the toucan, to ferret out and confront the evil spirits that have caused harm to his young patient.

The shaman identifies the evil spirit as a forest bird and begins to chant to it, challenging its sovereignty over the boy. In his vision, the shaman reaches for his bow and arrow and, aiming precisely, shoots the arrow into the bird spirit,

A Waiká Indian inhales from a wooden tube through which his companion is blowing *ebena* snuff (produced from *Virola* species). In less than a minute, the bioactive compounds are introduced into the man's bloodstream, and, after a few minutes, they induce powerful hallucinations.

Shamanistic Healing:
Fact or Illusion?

One study of shamanistic healing observed under clinical conditions oc-curred in the Orinoco Valley of Colombia. A Guahibo male, aged 24 years, was admitted to a medical clinic 24 hours after having been bitten by a *Bothrops*, a venomous snake. The patient was pale, confused, and delirious. Blood pressure was low at 90/50, pulse 100, respiratory frequency 32, and temperature 36.2°C. The patient exhibited severe edema and hyperthermia with cyanosis (purplish discoloration of the skin). There was a significant amount of blood in his urine. Antivenom serum was administered. Thirty minutes later, as the patient's condition worsened, a Guahibo shaman who was also present asked permission to administer a traditional "smoke-blowing" treatment. After lighting some tobacco, he began to blow smoke on the patient's extremities while intoning a monotonous chant similar to the song of a nocturnal bird. Within an hour the patient became relaxed and his vital signs returned to normal, despite the fact that from a medical perspective he remained in a toxic state. Over the next few days his general condition improved. In the attending physician's experience, antivenom serum alone had never produced such a dramatic effect, suggesting that a synergistic relationship existed between the Western and traditional treatments. Magnus Zethelius and Michael Balick, the authors of this study, concluded that the patient's recovery was enhanced by his strong belief and trust in traditional shamanistic practices.

---

causing it to release its grip on the boy. The profound effects of *ebena* on the chemistry of the brain transport the shaman into the world of his dream vision. An hour or so after the shaman has ingested the snuff, the vision comes to a close, and he takes a seat beside the boy and declares him healed.

Traditionally the ethnobotanist who observes such encounters has made no effort to diagnose the disease or verify the efficacy of the healer's treatment in Western terms; the investigator's role has been to document and collect the plants the shaman uses and to understand the cultural context of their use. Recently, however, increasing numbers of physicians have collaborated with ethnobotanists in fieldwork. Working as a team, they can analyze healing treatments from both Western and indigenous perspectives.

Psychoactive plants are an integral part of the traditional healing practices of many indigenous groups in parts of Africa and throughout the Neotropics. The Waiká shaman's snuff is a composite of three plants, each of which contributes compounds that enhance the bioactive power of the others. One of the components, *Virola theiodora* [Myristicaceae], was first described by Richard Spruce in 1851 during his explorations in the forest near Manaus, Brazil. At the time, however, he did not realize that it was used in hallucinogenic snuff.

Because the active molecules in hallucinogenic snuffs are structurally complex and deteriorate rapidly under tropical field conditions, it has been extraordinarily difficult to identify the chemical components of these plant mixtures precisely. It is even more difficult to understand how the human body metabolizes the compounds of fresh psychoactive herbs. Students of indigenous psychoactive plants have often wished they could somehow transport the shaman and his or her plants to a well-equipped laboratory for precise analysis. Frustrated in that desire, Richard Evans Schultes did the next-best thing: he took a well-equipped laboratory to the shaman.

In 1977 the research ship *Alpha Helix,* staffed with chemists, pharmacologists, and ethnobotanists, made its way up the Ampiyacu River, a tributary of the Amazon near the border between Peru and Colombia. On board were the ethnobotanists Richard Evans Schultes and Timothy Plowman, the pharmacologists Bo Holmstedt and Laurent Rivier, and the chemist Neal Towers. This expedition, sponsored by the U.S. National Science Foundation, sought to explore the pharmacological basis of toxic and medicinal plants used by the Indians of the Amazon Basin. The scientists collected an array of plants and brought them back to the ship, where they were analyzed with a tool most ethnobotanists only dream of having access to in the field—a gas chromatograph. Reading the graphical output of the gas chromatograph, the scientists were able to rapidly determine the specific compounds present in the sample. One project involved analysis of the sap of *Virola* trees, used as a hallucinogen. Another analyzed the presence of cocaine in the blood after the ingestion of coca leaves. A major advantage of having the gas chromatograph on the ship, only a few hundred meters away from the plants and the people preparing them, was that the material could be analyzed in exactly the state in which the local people ingested it. The careful chemical analyses carried out on the *Alpha Helix* and by other interdisciplinary collaborators have begun to reveal the molecular bases of plant hallucinogens.

To make *ebena* snuff, the shaman first heats bark peeled from a species of *Virola* tree carefully over a fire, releasing red sap. He then grinds dried leaves of *Justicia pectoralis* in a small pot. The resin and dried leaves are mixed with an ash from *Elizabetha princeps,* and the mixture is kneaded until it has a puttylike consistency. The mixture is then heated and ground to powder. Finally, the fine powder is poured onto a palm leaf any extraneous material is removed, and the powder is stored until use.

(1)

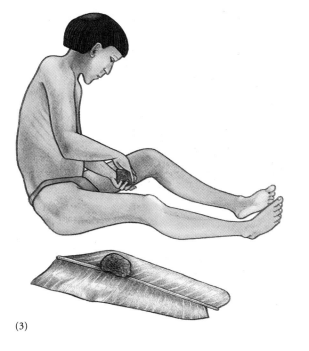

(3)

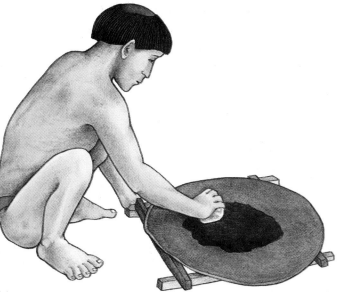

(4)

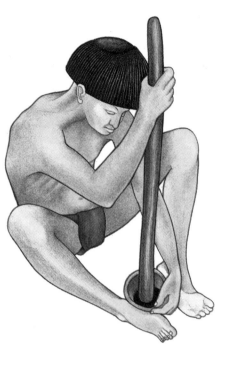

(2)

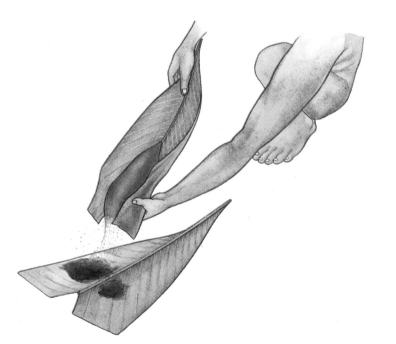

(5)

Virola sap has been found to contain such powerful psychoactive compounds as tryptamines, particularly N,N-dimethyltryptamine (DMT), N-monomethyltryptamine (MMT), and 5-methoxy-N,N-dimethyltryptamine (5-MeO-DMT), which are chemically related to 5-hydroxytryptamine (serotonin), which is found in the human brain. *Virola* also contains various beta-carbolines, compounds that enhance the effects of orally administered tryptamines and are psychoactive in their own right as well. The Waiká may not know the chemical composition of *ebena* snuff, but they have discovered how to prepare it for maximum potency. The Waiká use only fresh material. Removing the tree's outer bark, the shaman strips off the inner bark in pieces of 50 centimeters by 5 centimeters and wraps them in a large leaf for transport back to the village. He heats the strips over a fire, releasing a red sap, which he then mixes with two other plant materials, the ash of the tree *Elizabetha princeps* [Caesalpiniaceae] and the dried, powdered leaves of the herb *Justicia pectoralis* [Acanthaceae]. The ethnobotanist Peter de Smet of the Royal Dutch Association for the Advancement of Pharmacy in The Hague suggests that calcium carbonate crystals in the cells of *J. pectoralis* may facilitate the extraction of several tryptamine alkaloids from *Virola* and their absorption through the mucous membranes. The selection of these three plants, which exert such powerful synergistic action, is striking testimony to the skills of the shaman who made this mixture from forest plants and certainly to the abilities of his predecessors who first discovered and compounded this mixture.

The snuff must be prepared carefully, for any error will render it lethal or less effective. *Ebena* is so potent that it can kill an older shaman even when it is prepared properly. The shaman kneads the three plants together between his legs till they have the consistency of putty. He then toasts the mixture, and when it is hard he grinds it into a powder. He carefully places the powder on a *Geonoma* [Arecaceae] palm leaf, which protects it from the elements so it will be clean and usable for the next day's ceremony.

How do traditional peoples discover that certain plants produce useful pharmacological benefits? How do they discover a means of extracting bioactive fractions that are useful as well as safe? Over many generations of living in and relying on the natural environment, they have had myriad opportunities for experimentation. Indigenous peoples' science, like our own, proceeds through a process of trial and error during which it encounters both errors and successes, then builds on these successes and failures in an iterative pattern of observation upon observation. The ultimate results of this hypothesis testing are astounding when one considers that some 250,000 flowering plants are found on earth. But we humans are consummate experimenters, driven by curiosity and the prospect of finding a more pleasant, adventurous, or meaningful life. The skills displayed by the indigenous groups of the Amazon Basin, who have discovered how to use complex chemical admixtures of plants with potent psychoactive effects, are in some ways analogous to those of a chemist or pharmacologist.

Traditional peoples are sometimes aided in their search for new substances by the classification systems they have developed. For example, groups such as the Barasana Indians of the Northwest Amazon Valley of Colombia distinguish different forms of the psychoactive plant *Banisteriopsis caapi* [Malpighiaceae] on the basis of the colors or types of visions they produce. One form can produce red visions; another form produces visions that include people; still another form will make one fierce and strong. Similarly, according to Richard Evans Schultes, the Indians in this area recognize some 14 forms of the stimulant plant *Paullinia yoco* [Sapindaceae].

Such ethnotaxonomic systems are the basis for experimentation in the search for additional useful species. People may experiment with a plant that looks similar to a species with a known use, such as a food plant, hoping to find superior properties in the new species. This trial-and-error process is more firmly based on logic than the term suggests.

Of particular interest is the stunning chemical similarity among hallucinogens produced from widely disparate plants and used by widely disparate cultures. Even more interesting are the structural similarities between some of these plant chemicals and compounds that occur naturally in the human brain. The resultant sophistication in both pharmacology and psychoactive effect can best be seen in a South American plant, ayahuasca.

Ayahuasca, Vine of the Soul

The Quechua Indians of Ecuador say that ayahuasca ("vine of the soul") has the ability to release the spirit and allow it to wander freely before returning to the body. The earliest report of the psychoactive properties of this vine was provided by M. Villavicencio in *Geografía de la República del Ecuador* in 1858. He noted that the tribes of the Río Napo drank a beverage prepared from ayahuasca to

> foresee and to answer accurately in difficult cases, be it to reply opportunely to ambassadors from other tribes in a question of war; to decipher plans of the enemy through the medium of this magic drink and take proper steps for attack and defense; to ascertain, when a relative is sick, what sorcerer has put a curse; to carry out a friendly visit to other tribes; to welcome foreign travelers or, at last, to make sure of the love of their womenfolk.

Banisteriopsis caapi, known as *caapi,* ayahuasca, or *yajé,* is a vine native to the Amazon region of South America. By boiling its stem in water, traditional peoples of the Amazon prepare a hallucinogenic beverage used for divination and telepathy. Ayahuasca is a Quechua term translating as "vine of the soul," meaning that the plant releases the spirit of the person who consumes it.

Although traditional cultures in many parts of the world have employed plants for their hallucinogenic properties, those of the Western Hemisphere have exploited a far greater variety of species for that purpose. The map shows the regions in which selected plants were used as hallucinogens.

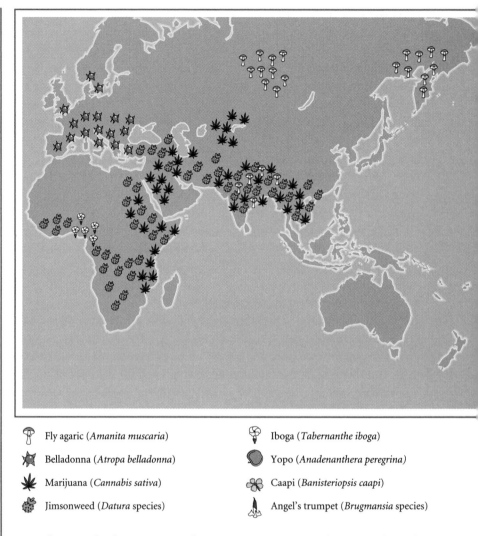

🍄 Fly agaric (*Amanita muscaria*)		🌿 Iboga (*Tabernanthe iboga*)	
✳ Belladonna (*Atropa belladonna*)		⬤ Yopo (*Anadenanthera peregrina*)	
🌿 Marijuana (*Cannabis sativa*)		❀ Caapi (*Banisteriopsis caapi*)	
🌸 Jimsonweed (*Datura* species)		🌼 Angel's trumpet (*Brugmansia* species)	

Ayahuasca, also known as *yajé*, has a strange reputation that stems from the plant's purported abilities to enable telepathy and foretell the future. Two species of *Banisteriopsis* are of primary importance, *B. caapi* and *B. inebrians*. The plant is prepared by cutting fresh sections of the stem and then boiling the bark in a pot of water for a few hours. The decoction is bitter, and only a small amount is consumed. Many tribes in South America use it as part of an elaborate ritual that may include dancing, chanting, and other activities. The psychoactive effects of ayahuasca are due to beta-carboline alkaloids such as harmine and harmaline, along with other minor alkaloids. Admixtures of various plants that contain tryptamine derivatives, such as *Banisteriopsis rusbyana* and *Psychotria viridis* enhance the effects, although ayahuasca is powerful

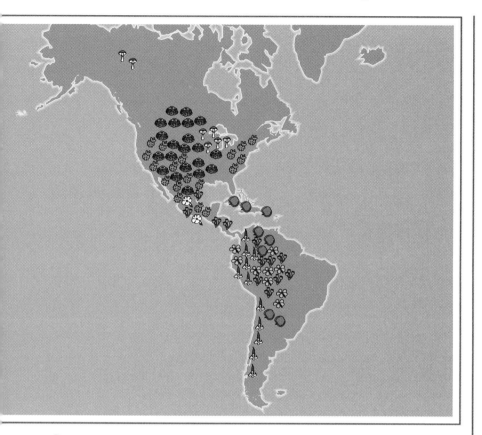

Peyote (*Lophophora williamsii*)

Hallucinogenic mushrooms (*Psilocybe, Conocybe, Panaeolus, Stropharia* species)

Morning glory *(Turbina corymbosa* and *Ipomoea violacea)*

Ebena *(Virola theiodora)*

enough to be used alone. The leaves of *P. viridis* contain 0.1 to 0.5 percent DMT (N,N-dimethyltryptamine), a powerful hallucinogen that in itself does not produce a hallucinogenic effect when it is taken orally because monoamine oxidase (MAO) inhibitors in the liver and stomach inactivate it. The beta carboline alkaloids in the bark of *B. caapi,* however, inhibit MAO and allow the DMT to cross the blood-brain barrier without being degraded. Similar monoamine oxidase inhibitors have been found to have a direct effect on brain chemistry and are used to treat depression.

The structural similarity of some molecules found in hallucinogenic mixtures to 5-hydroxytryptamine (serotonin), a major chemical messenger in the brain, is staggering. All of the chemical structures on the following page—

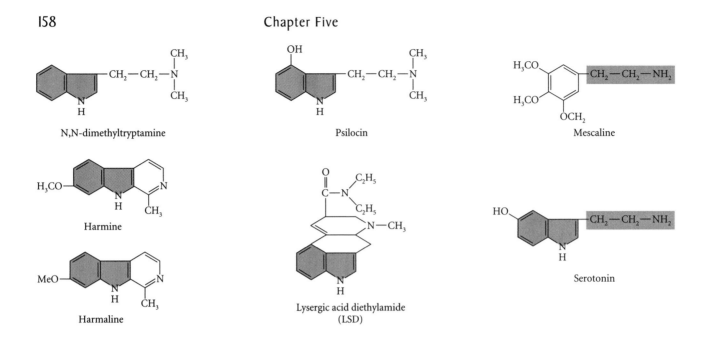

N,N-dimethyltryptamine

Psilocin

Mescaline

Harmine

Harmaline

Lysergic acid diethylamide
(LSD)

Serotonin

Many of the alkaloids responsible for hallucinogenic activity share the indole nucleus consisting of phenyl and pyrrol segments (colored portions of the molecules). The indole nucleus is also present in serotonin, a powerful chemical messenger found in the brain. Though lacking the indole nucleus, mescaline has a side chain identical to that of serotonin. Serotonin plays an important role in mediating mood and emotion.

N,N-dimethyltryptamine in *ebena* snuff or ayahuasca, the beta carbolines harmine and harmaline in ayahuasca, the psilocin in the hallucinogenic mushrooms *Psilocybe, Conocybe, Panaeolus,* and *Stropharia* and LSD-25, a synthetic derivative of a substance produced by the ergot fungus—have the same core chemical structure as serotonin. Other psychoactive molecules mimic the side chain of serotonin: the side chain of mescaline (from the peyote cactus) is very similar to that of serotonin. Despite the structural similarity of the psychoactive molecules to serotonin, their mode of action is not known. It has been proposed that contamination of food grains by the ergot fungus (which produces LSD) may have contributed to the onset of the Dark Ages in Europe.

The ayahuasca drink, like many psychoactive preparations, may be made in many ways. Various admixtures influence the dosage, administration, and effects of the drug. Schultes commented on his own experience with ayahuasca:

> The intoxication began with a feeling of giddiness and nervousness, soon followed by nausea, occasional vomiting and profuse perspiration. Occasionally, the vision was disturbed by flashes of light and, upon closing the eyes, a bluish haze sometimes appeared. A period of abnormal lassitude then set in during which colours increased in intensity. Sooner or later a deep sleep interrupted by dream-like sequences began. The only uncomfortable after-effect noted was in-

testinal upset and diarrhoea on the following day. At no time was movement of the limbs adversely affected. In fact, amongst many Amazonian Indians, dancing forms part of the caapi-ritual.

Ayahuasca is one of the most powerful tools of the Amazonian shaman; but he is not the only one to use it. Many other members of the community, Schultes notes, drink it "to see all the gods, the first human beings, and animals, and [to] come to understand the establishment of their social order."

Some modern Peruvians use the plant to help preserve their traditional ways. Pablo Amaringo, a Peruvian shaman well versed in the use of ayahuasca, has produced paintings of the visions he sees under its influence. Amaringo's paintings are known for their rich and vivid color, symbolism, and detail. He founded a school for Amazonian artists, the USKO-AYAR Amazonian School of Painting in Pucallpa, Peru, and is currently training young artists to use this powerful medium to record and preserve their heritage. The carefully controlled use of ayahuasca is embedded in the native religion, but its use outside of its traditional context can be dangerous. Ayahuasca has become a recreational drug in some Amazonian towns: some tourists are attracted by the possibility of a shamanistic experience, while some local people use it to divine the future or to manage their personal lives (such as checking up on the faithfulness of a spouse). There is considerable reason to believe that the use of hallucinogenic plants outside of their traditional religious contexts can produce sorrow rather than transcendence, confusion rather than enlightenment.

Lewin's Categories

The hallucinogenic effects of *ebena* snuff and ayahuasca are typical of one type of psychoactive plant; other plants alter mood or perception more gently. In an effort to systematize the various types of psychoactive plants, the German toxicologist Louis Lewin in 1924 published a book that appeared in English in 1931 as *Phantastica: Narcotic and Stimulating Drugs—Their Use and Abuse*. In it Lewin examined the botany, ethnobotany, ethnology, chemistry, history, and pharmacology of psychoactive substances, and elaborated on their place in indigenous cultures and in modern medicine, psychology, psychiatry, and sociology. He discussed 28 plants as well as a number of synthetic compounds "capable of effecting a modification of the cerebral functions, and used to obtain at will agreeable sensations of excitement or peace" by people around the world, from the great cities of Europe to small villages in tropical forests.

Lewin proposed five categories of narcotic and stimulating plants. The first group he called "euphorica," or "sedatives of mental activity." Included in this group are opium and its derivatives morphine and heroin, as well as cocaine. These drugs lower or suspend "the functions of emotion and perception in their widest sense, sometimes reducing or suppressing, sometimes conserving consciousness, inducing in the person concerned a state of physical or mental comfort." The second group, "phantastica," includes drugs that cause "evident cerebral excitation in the form of hallucinations, illusions, and visions . . . [which] may be accompanied or followed by unconsciousness or other symptoms of altered cerebral functioning." The third category, "inebriantia," consists of substances produced primarily by chemical synthesis or some other manipulation of raw materials (such as alcohol, chloroform, ether, benzene) to induce "a primary phase of cerebral excitation . . . followed by a state of depression." In the fourth group, "hypnotica," Lewin placed substances that induce sleep, such as kava, used in the Pacific islands. The final group, "excitantia," consists of stimulants that result in "apparent excitation of the brain" without altering consciousness. Among such substances are coffee, tobacco, betel nuts, and cola nuts.

Lewin hoped that his work would lead to further investigations, and indeed it did. As Richard Evans Schultes later wrote, "We may truly say that it was Lewin's *Phantastika* that sparked to-day's interest in narcotics, especially in those that we have come to refer to as the hallucinogens." Lewin's classification system is still used to teach students of ethnobotany and pharmacology about these useful and harmful biodynamic substances.

Kava and Psychoactive Drugs as a Communal Experience

A hallucinogen such as *ebena* snuff or ayahuasca produces such a powerful experience that indigenous peoples believe their souls are transported to another world. Other types of psychoactive plants, however, are far more subtle and appear to facilitate social interactions in this world. Included in this group are the plants that yield such recreational beverages as wine, maté, coffee, chocolate, and tea, as well as the tobacco used in Native Americans' peace pipes. People on the islands of the South Pacific prepare a beverage called kava from the rhizomes of *Piper methysticum* [Piperaceae]. Though kava consumption can have soothing and sedating effects that make it useful to induce sleep or give comfort to the ill, its principal social use is to build community and avoid conflict. The Polynesians ceremonially drink kava to welcome visitors to their villages and to help the villagers reach consensus on potentially controversial decisions affecting the community.

The roots and rhizomes of the kava plant *Piper methysticum* are used throughout the islands of the South Pacific to prepare a ceremonial beverage that helps establish feelings of fellowship and amicability among villagers.

The ethnobotanists Vincent Lebot at the University of Hawaii and Pierre Cabalion at the Herbarium, Office de la Recherche Scientifique et Technique d'Outre Mer in Vanuatu have recorded 15 different kava lactones in kava roots. Kava lactones consist of 13-carbon molecules attached to a lactone, a cyclical carbon ring with a double-bonded oxygen. The most powerful of these— kavain, dihydrokavain (DHK), and dihydromethysticin (DHM)—have a modest analgesic effect, about twice that of aspirin, and act as a mild anaesthetic and tranquilizer. The effects of drinking kava are subtle and nuanced: one feels tranquil, though the mind remains extraordinarily clear. Lebot has identified nine major groups of kava clones. Each has a slightly different proportion of six major kava lactones, and, consequently, each produces a slightly different psychoactive effect. A Samoan clone, called "fellowship and brotherhood," makes one feel very friendly. Another type, called "the white pigeon," imparts a sense of heightened perception, as though one were flying over the rain forest like a pigeon. Lebot finds that as the islanders transported kava from Melanesia to Polynesia, they selected plants poor in dihydrokavain and rich in kavain.

The slight tranquilizing effect of kava makes it an ideal beverage to serve when vexatious matters such as land disputes are to be discussed or at times of apprehension, as when strangers appear. Both the ceremony and the beverage

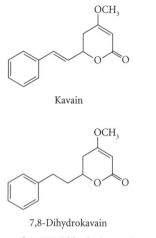

Kavain

7,8-Dihydrokavain

Kavain and DHK (dihydrokavain) are the two kava lactones that most easily pass into the brain through the blood-brain barrier. Although the two lactones differ by only a single chemical bond, they have different physiological effects. Kavain acts in the mouth as a local anesthetic, producing effects like those of cocaine, but is less easily absorbed and metabolized than DHK. As the original colonists transported kava eastward into Polynesia, they selected for plants rich in kavain but poor in DHK.

seem designed to increase friendly feelings and reduce the possibility of hostility. Throughout the islands of Polynesia and many islands of Melanesia kava is the symbol of friendship. In most Polynesian cultures, the serving of kava is the highest honor accorded a visitor. Visitors who drink kava with village chiefs can be assured of the villagers' friendship and hospitality for the duration of their stay. The real power of kava comes from the cultural context in which it is drunk. Consumed in one's home, kava has an effect that is scarcely noticeable. But drinking kava under a thatched roof 10 meters high in the presence of the assembled chiefs of the entire district, all of whom scrupulously follow the ancient forms of rhetoric, is a truly memorable experience.

In Samoa the kava ceremony is a beautiful and emotional event. In formality it can be likened to the Japanese tea ceremony. In eloquence it easily exceeds even the most stirring speech in Parliament or Congress. In sacredness it can be compared to the mass; just as the mass was long conducted in Latin, so the kava ceremony is conducted in the formal Samoan language of respect. The kava ceremony, then, is not easily accessible to foreigners, even to those fluent in the common language of Samoa. But even a slight knowledge of Samoan respect language yields great rewards, for it enables one to grasp the beauty of the kava ceremony. "Our meeting is as the tips of two clouds passing in the sky," the chief says at the beginning of the ceremony:

Ua mamalu lo tatou taeao fesilasilafa'i e pe o le ta'otoga o i'a sa. Ua pa'ū le vao, ua liligo le taeao e pe o le ulua'i ave o le la ua fa'asuluina le fogā'ele'ele. Ua mamalu le taeao e pei ua fa'afesiligia le fogāmauga e le fogāsami pe aisea e tulu'i ai loimata e pei o le timuga mai le lagi. E fia fa'afofoga mauga o Salafai; e fia fa'alogo galu o le sami, auā lenei taeao lalelei. Ua pa'ia le sami, ua pa'ia le fanua, ua pa'ia le malae, ua pa'ia le maota, ma ua matou tau pa'i malu atu i le pa'ia ma le mamalu o le au fa'afofoga mai. . .

Our meeting is as the mating of sea turtles, silent, motionless, but sacred. Our meeting is as sacred as the first dew, as sacred as the first ray of light that filled the newly created earth. Our meeting is as sacred as the meeting of the sea and mountains, who looked upon the sun and asked why it wept tears of rain from the heavens. The very mountains and the waves of the sea are moved at our meeting on this morning. The sea is sacred, the earth is sacred, our meeting ground is sacred, our meeting house is sacred, and it is with trembling that we address the sacredness and dignity of those who listen. . . .

The drinking of kava links the people present with generations past, back to the beginning of time. In the village of Fitiuta orators still tell the story of the first kava ceremony. Tagaloalagi, who created the world, had a kava ceremony with the first man, Pava. The space between the two parties facing each other in a kava ceremony, called the *alofi*, is most sacred. Once a kava ceremony begins, it is forbidden to stand, walk, or intrude in any fashion into the *alofi*. During the ceremony with Tagaloalagi, Pava's son stepped into the sacred space. "Forbid your son," Tagaloalagi commanded Pava. "He must not violate the sanctity of the *alofi*."

But the young boy ran between Tagaloalagi and Pava, so Tagaloalagi reached over, grasped the boy, and ripped him limb from limb. Pava began to weep at the death of his only son, the only hope to populate the world. "Your son violated the *alofi* and so had to die," Tagaloalagi said as he lifted the kava cup. "But though through transgression came death, through kava comes life."

Kava ceremonies rival the Japanese tea ceremony for intricacy of action and sophistication of rhetoric. In Samoa the *taupou* or village virgin was traditionally the only person allowed to prepare and serve kava to assembled chiefs on ceremonial occasions. This custom continues in many villages today.

Tagaloalagi dribbled out a few drops of kava on the dismembered body, and the boy was resurrected on the spot. "And the sacred kava will always serve as a covenant between you and me." Overjoyed to see his son whole again, Pava clapped his hands. And then he and Tagaloalagi drank the sacred beverage.

Today during kava ceremonies throughout the Pacific, the participants clap their hands for joy and dribble a few drops on the mat before they drink. Only a few knowledgeable old orators remember why, but the power and beauty of the allegory of the Fall and Resurrection remains.

Cannabis in World History

Not all psychoactive plants were used to promote social tranquility. Even the sense of community created with the aid of psychoactive plants can be used for

Archaeologists have found various objects employed during *Cannabis* rituals in Scythian tombs, in the Altai Mountains. The Scythians stored *Cannabis* fruits in pots like that at the lower right and burned *Cannabis* in copper vessels like that at the upper right to produce intoxicating smoke. The vessel with the smoldering *Cannabis* was placed inside a small tent, about 45 centimeters tall, formed by several poles (left) covered with animal skins, from which someone could inhale vapors of this psychoactive plant.

darker purposes. One example comes from the use of marijuana in the eleventh century. Al-Hasan ibn al-Sabbah, a Persian by birth, established an enclave overlooking a strategic caravan route near Baghdad. By 1090 he had broken away from the traditional Muslim teachings of the day and founded his own sect, whose practitioners were known as *hashishin*, after the full name of al-Hasan. They lived off the riches they obtained by robbing caravans on the road to Baghdad and built beautiful palaces and gardens for their followers. Soon the *hashishin* numbered over 12,000, attracting many young men into the cult who were then schooled in the arts of robbery and assassination. Travelers to the court of al-Hasan described it as an earthly paradise where many young men, sedated by a beverage derived from *Cannabis* [Cannabaceae], lived in bliss, their every need satisfied. Al-Hasan employed these men to spread his religion, increase his wealth, and destroy those who opposed him. If they should die in carrying out their mission, he told them, the reward for their heroism would be eternity in paradise, where they would live much as they did now in al-Hasan's palaces. His soldiers, known as "assassins" (a corruption of *hashishin*), were fierce, determined, and extraordinarily successful, for they believed that an eternal reward awaited them. But use of *Cannabis* far predates the eleventh century.

The writings of the Chinese emperor Shen Nung, traditionally dated before 2000 B.C., mention *Cannabis* as an important plant for the treatment of various human illnesses, including beri-beri, malaria, and forgetfulness. The emperor also described it as a plant that frees the psyche: "If taken over a long term, it makes one communicate with spirits and lightens one's body." Later writers in China, however, warned that it was a "liberator of sin." The *Rh-Ya*, compiled during the fifteenth century B.C., contains the earliest mention of this plant, known as *ma*, for shamanistic purposes. By the second century A.D. Chinese physicians mixed it with wine and gave it to patients before surgery, as an anaesthetic, for it was said to dull pain. Opium, however, was the drug of choice for this purpose.

The discovery near Jerusalem of *Cannabis* in the abdominal cavity of the skeletal remains of a young woman who apparently died in childbirth during the fourth century B.C. has led to speculation on the use of this drug in the ancient Middle East. The authors of this study concluded that the drug was administered as an inhalant to reduce pain and increase the force of uterine contractions. So far there have been no similar archaeological findings, so it is difficult to know whether this was a widespread practice.

More ritualistic uses of *Cannabis* as a psychotropic drug appeared in Asia Minor among the ancient Scythians. Archaeological excavations have uncovered pots and charcoal containing the remains of *Cannabis* leaves and fruits dating between 500 and 300 B.C. According to the Greek historian Herodotus,

This bag of *Cannabis* leaves that has been gathered for medicinal use in Nepal. *Cannabis sativa* is used therapeutically in a number of traditional medical systems, including the Ayurveda system that is widely practiced in Nepal and throughout the Indian subcontinent. *Ganja* or *bhang*, as the plant is known in Hindi, is traditionally used as an antispasmodic, analgesic, and sedative.

the Scythians delighted in vapor baths scented by *Cannabis* seeds placed on heated stones. Ethnobotanist William Emboden concluded that among the Scythians, as well as other people in this region, shamanistic ritual with *Cannabis* played an important role in curing the sick and in sending the deceased into the next world.

Much later there was a surge of interest in *Cannabis* among the avant-garde community in Paris. When Napoleon's army returned to France after its Egyptian campaign around 1800, his troops brought with them a new custom, the consumption of *Cannabis* resin. At first they used it to treat the mentally ill, but the focus changed from therapy to recreation when some fashionable Parisians took to using it when they got together. The group so thoroughly enjoyed their shared visions that in 1844 they formalized their gatherings as Le Club des Haschischins and held monthly meetings at the Hôtel Pinodan on the Île Saint-Louis. After they imbibed the intoxicating resin, they were served an elegant meal; then the effects of the hallucinogenic substance would become apparent. Thus began the popular use of *Cannabis* in Europe.

Despite all efforts to eradicate it, the consumption of *Cannabis* in its various forms has spread throughout the world. Its powerful resins contain delta-1-tetrahydrocannabinol, the most active form of the more than 30 cannabinoids found in the plant. This drug exerts a powerful effect on the central nervous system, causing euphoria and changes in perceptions of space, time, and taste, as well as other effects. The leaves and flowers are harvested, dried, and smoked in cigarette form or in a pipe. Hashish, made from resin derived from the pistillate flowers, is a much more powerful form of the drug. Millions of people use it in the Middle East and Africa, and elsewhere. The plant and its products can be harvested in many ways; at one time men ran naked through *Cannabis* fields in Nepal and later scraped the sticky resins from their bodies.

Europeans' fascination with psychoactive drugs, perhaps spurred by early experience with *Cannabis,* led them to experiment with a variety of other plants. Plants that once were the focus of sacred rites now produce compounds irreverently used as recreational substances. For the most part, Westerners have ignored the traditional cultural and religious contexts in which these plants were used, leaving the way open for some, including those Western scientists lacking ethnobotanical data, to demonize both the plants and their uses. In the 1890s, for example, Emil Kraepelin, a German psychiatrist, equated the Andean Indians' practice of chewing coca leaves with the abuse of cocaine and suggested that the consequences of both were equally tragic. The story of coca is one of the most pronounced examples of the denigration of psychoactive plants that are so important to traditional cultures.

Coca and Cocaine

The coca bush, an indigenous plant of South America, is the source of the alkaloid cocaine and a derivative, cocaine hydrochloride. It is important to distinguish coca leaf from cocaine. The plant that yields the leaves is highly esteemed and has been used for centuries by people of the Andes and the Amazon region to allay hunger and fatigue, as a medicament, and as a source of nutrients. Cocaine, a powerful and addictive substance derived through chemical treatment

A ceramic vessel depicting a coca leaf chewer, from La Libertad, Peru, dated ca. A.D. 400–600. Archaeological finds such as this one show the antiquity of coca chewing, and its importance to the Indian cultures of South America.

of the leaf, induces cerebral stimulation and euphoria. It can both reduce pain and inflict suffering, depending on dosage and form and on whether it is used for the treatment of illness or for recreational purposes. Cocaine hydrochloride is used in modern medicine as a local topical anaesthetic applied to mucous membranes, and also in a mixture of drugs to control the pain of terminal cancer.

The coca bush is in the genus *Erythroxylum* [Erythroxylaceae]. In over a decade of fieldwork, the late Timothy Plowman, without doubt the most gifted

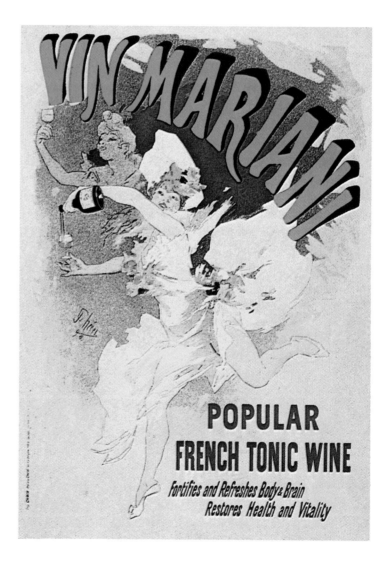

This poster, produced in 1894, advertised the virtues of Vin Mariani, a popular wine made from coca leaves.

student of Richard Evans Schultes, found four varieties to be the source of all cultivated coca: *Erythroxylum coca* var. *coca* (Huanuco or Bolivian coca), *E. coca* var. *ipadu* (Amazonian coca), *E. novogranatense* var. *novogranatense* (Colombian coca), and *E. novogranatense* var. *truxillense* (Trujillo coca).

Coca was domesticated in pre-Columbian times and has become intricately woven into the fabric of the indigenous cultures in the regions where it grows. The early Spanish chroniclers who visited the Andean regions noted that the Indians there chewed the dried green leaves with mineral lime, a substance that helps the mucous membranes in the mouth absorb the alkaloids in the leaves. This practice continues to this day. Archaeologists have recovered ceramic vessels that represent people with one bulging cheek, a clear indication of the presence of a quid of the lime-leaf mixture. Plowman speculated that *Erythroxylum coca* var. *coca* was brought under cultivation some 7000 years ago in the eastern Andes. He believed that Amazonian coca, cultivated in the western Amazon region, was domesticated much more recently.

Accolades for Mariani Wine

"My conversion is complete. Praise to Mariani's wine!"—Zadoc Kahn, Grand Rabbi of France

"Your coca from America gave my European priests the strength to civilise Asia and Africa."—Cardinal Lavigerie

"Mariani, your sweet flasks delight my mouth."—Alexandre Dumas fils

"Since a single bottle of Mariani's extraordinary coca wine guarantees a lifetime of a hundred years, I shall be obliged to live until the year 2700!"—Jules Verne

"To Mariani, who spreads coca."—Auguste Rodin

Mariani also received letters of appreciation from hundreds of others, including President William McKinley of the United States, Prince Albert of Monaco, and King Alphonse XIII of Spain.

These bundles of coca leaves are being sold in the outdoor market in Las Paz, Bolivia, where use of the leaf as a masticatory and medicament is common.

Small ceramic containers for lime, typically associated with coca chewing, found at an archaeological site in coastal Ecuador indicate that this plant has been in use there for at least 5000 years. Today on the streets of La Paz, Bolivia, it is common to see vendors with huge bales of coca leaves for sale. In this area people chew the whole leaf, followed by lime from a special gourd container. A visitor unaccustomed to the high altitudes of the Andes has only to drink a cup of hot coca tea to obtain prompt relief from the headache, nausea, and weakness of *soroche*, or altitude sickness, and the fatigue of travel. Coca leaf also contains an assortment of vitamins and minerals that make a valuable contribution to the often impoverished local diet. One hundred grams of Bolivian coca leaves contain more calcium, iron, phosphorus, vitamin A, vitamin B_2, and vitamin E than are found in the U.S. recommended daily dietary allowance. The calcium content is 1540 milligrams per 100 grams of leaf, an important nutritional supplement in a region where the diet is frequently low in milk products. Many contemporary entrepreneurs in Bolivia promote the use of products containing coca leaf extracts—toothpaste, chewing gum, and a variety of tinctures, wines, and liqueurs. Europeans have added cocaine to various alcoholic beverages, as well.

Angelo Mariani, an Italian physician who lived in France, produced a red Bordeaux wine mixed with an extract of coca leaves between 1844 and 1913. Known as Vin Mariani, the beverage was produced in Paris, dispensed by prescription, and distributed throughout the world. Pope Leo XIII presented Mariani with a gold medal as a token of the esteem in which he held the product. Mariani was so proud of the letters of appreciation and endorsement he received that he compiled and distributed them. They filled 13 volumes.

Coca was not used solely as an additive to wine. An American pharmacist, John Styth Pemberton, used coca leaf to flavor a beverage he concocted, which he called Coca-Cola. As Pemberton originally formulated it in 1886, Coca-Cola was intended to be a "brain tonic and intellectual beverage"; in addition to cocaine, it contained caffeine derived from African cola nuts. A federal law of 1904 required cocaine to be removed from Coca-Cola, but the leaf extract is still included under the heading of "natural flavors." During the 1980s, the Stepan Company of Maywood, New Jersey, imported as few as 56 and as many as 588 metric tons of leaf per year for use in Coca-Cola. The cocaine-containing extract that Stepan removes is sold to Mallinckrodt, Inc., which purifies it into cocaine hydrochloride for use in the pharmaceutical industry as a local anaesthetic.

But there is an extraordinarily dark side to the trade in coca leaves. In the interior markets of Bolivia or Peru, anyone can purchase a hundred kilograms of coca leaves for approximately $66. Using simple techniques the purchaser can process these leaves into 2 kilograms of crude cocaine paste, which, with further processing, yields approximately 1.5 kilograms of pure cocaine. A kilogram of pure cocaine can be purchased, albeit illegally, in Bolivia, Peru, or Colombia for anywhere from $1500 to $2500. That kilogram of cocaine, however, becomes much more valuable if it is transported to New York City, where it would wholesale at $25,000 to $35,000. If the dealer then breaks the kilogram into small lots, its retail value soars to more than $100,000 — more than 1500 times the value of the unprocessed leaves. To achieve these massive profits, smugglers are willing to brave almost any risk or engage in any form of violence. Indeed, the political scientist Lamond Tullis, of Brigham Young University, argues in his recent book *Unintended Consequences* that governments who attempt to choke cocaine smuggling at the source only cause the price of cocaine to increase, providing even a greater profit and a greater incentive for would-be smugglers to accept large risks and engage in violence. Tullis fears that as long as there is an insatiable market for crack cocaine in the United States, smugglers will devise ways to service that market.

Since the 1960s the use of cocaine as crack and in other forms has exploded into a major social problem in the United States and other countries. In addition to its obvious impact on public health, the widespread sale and use of this drug has torn apart families, corrupted governments, and fomented violence and damage that may take generations to overcome. In this respect, the abuse of cocaine repeats a pattern established in an earlier generation by the abuse of another substance that can be traced to a psychoactive plant: heroin.

Opium and the Production of Heroin

The poppy from which opium is derived, *Papaver somniferum* [Papaveraceae], is not known to grow in the wild; it has been domesticated for its seeds, which are used for oil and food, and for its dried sap, which produces opium. For centuries people have bred poppies to increase the size of the capsules that contain the nutritious seeds and the opium-rich sap. They have also collected the tender leaves as an edible herb.

The ancient Greeks and Egyptians probably used opium poppies. A statue found in Crete, thought to be 3500 years old, appears to depict three poppy capsules in the headpiece of a smiling female figure with closed eyes. It has been suggested that she is a goddess in an opium trance. The Ebers Papyrus, an Egyptian compendium of medical information from 1500 B.C., mentions the opium poppy as a remedy for head pains and as a sedative. People of other ancient civilizations, such as that of Thailand, may also have used the plant. We cannot be sure of any of these interpretations, however, for images of pomegranates (*Punica granatum* [Punicaceae]) and water lilies (*Nymphaea caerulea* [Nymphaeaceae]) are sometimes remarkably similar to those of opium poppies.

A drawing of a 3500-year-old terra-cotta statue, discovered in Knossos, Crete, that is thought to depict a poppy goddess wearing a crown of three opium poppy capsules.

Once the poppy petals have fallen, opium is harvested from the green capsule. Horizontal slits are made in the capsule, usually with a multibladed knife. Care is taken not to cut through the capsule completely, for otherwise the sap will collect inside the capsule and the seeds cannot be harvested. The whitish sap is immediately secreted on the surface of the capsule and is collected with a curved scraper. The sap dries brown or black. It is kneaded into balls, which are dried in the sun to reduce the water content from 30 to 10 percent. The dried sap can be smoked directly for its effects. Commerce both legal and illegal centers on the refined alkaloids.

Opium contains more than 30 alkaloids, including codeine, morphine, noscapine, and papaverine. Codeine is an analgesic antitussive and is used in cough syrups. Morphine is a very powerful hypnotic and narcotic, with analgesic properties. It is the most abundant component of opium, ranging from 4 to 21 percent by weight. Noscapine, an alkaloid with no narcotic properties, is used to treat coughs. Papaverine is an antispasmodic and

Left: Heroin, synthetically modified from morphine by the addition of two acetyl groups (CH₃CO) was originally touted as a cure for morphine addiction. Although it produces fewer side effects (nausea and constipation) than morphine, it proved to be far more addictive. Right: The flower of the opium poppy, *Papaver somniferum*. The seed capsule above the flower has been sliced to release the sticky white sap that is harvested and dried to produce opium.

cerebral vasodilator. Paregoric, an opium tincture, is employed in the treatment of diarrhea. Heroin is a synthetic derivative produced by the acetylation of morphine—the replacement of the hydrogen atoms of both the alcoholic hydroxyl and phenolic groups with acetyl groups. Heroin, originally welcomed as a cure for opium addiction, proved to be more powerful than morphine and far more addictive.

Because of its potency and addictiveness, heroin is illegal in the United States and most other countries of the world. Legislation prohibiting traffic in psychoactive substances has been necessary to protect society from the ravages of drug abuse, but antidrug laws ignore the deeply religious uses of psychoactive plants by indigenous peoples. This conflict between indigenous religions and modern antidrug legislation is most apparent in the controversy concerning some Native Americans' use of peyote.

A peyote cactus in flower; note that a number of "heads" have developed above the single root. It is these heads that are cut from the root and dried into "buttons" for consumption as a hallucinogen.

Peyote and the Native American Church

Lophophora williamsii [Cactaceae], which Schultes considers a "prototype" of the hallucinogenic plants of the New World, was used perhaps as early as 2000 years ago. The Spanish conquistadores were astounded by the power of this "magic" cactus to induce brilliant visions, which seemed to allow the Indians to foretell the future. According to Fray Bernardo de Sahagún, who chronicled the life of the Indians of Mexico during the sixteenth century, "Those who eat or drink it see visions either frightful or laughable. This intoxication lasts two or three days and then ceases. It is a common food of the Chichimeca, for it sustains them and gives them courage to fight and not fear hunger nor thirst. And they say that it protects them from all danger." The Spaniards viewed peyote as an obstacle to their efforts to "civilize" the peoples of the New World, for the Indians used it in the practice of their own religion. Their efforts to eliminate that religion drove indigenous spiritual beliefs and practices into isolated regions, where they remain today.

The peyote cactus is native to central and northern Mexico and to the Rio Grande Valley of the southwestern United States. It is a small green-gray spineless plant with a deep taproot and often grows in groups. When the top is harvested, additional crowns often spring forth from the taproot, forming specimens with multiple heads. Among the peoples that use peyote are the Huichol and Tarahumara Indians of Mexico's Sierra Madre Occidental. These people undertake peyote hunts, often in early November, traveling to the areas where peyote is found, to gather it for use and distribution to tribes in other areas. The crowns can be dried into buttons that retain their potency for years. The buttons are taken into the mouth, moistened with saliva, and swallowed.

Mescaline, the alkaloid responsible for peyote's hallucinogenic property, accounts for some 30 percent of the total alkaloid content, although more than a dozen other alkaloids have been discovered. Ingestion of mescaline alone fails to give the multifaceted response of consuming the peyote cactus, so the other biologically active compounds in the plant appear to have some synergistic effects. The traditional dose of peyote ranges from 4 to 30 dried buttons; the intensity of visual hallucinations begins some three hours later. They are frequently accompanied by auditory, olfactory, and tactile hallucinations, a sensation of weightlessness, and an altered perception of time. Many indigenous peoples also consider peyote to be a medicine, for it facilitates contact with the spirits that cause certain illnesses.

In June 1887 John R. Briggs, a medical practitioner in Texas, sent peyote to Parke Davis & Co. in Detroit. The company's chemists discovered quantities of alkaloids in it. This was the first report of the occurrence of this class of chemicals in this plant family. Louis Lewin also published a report on the presence of alkaloids in the peyote cactus in 1888, and Parke Davis & Co. introduced a tincture of peyote in 1889. The tincture was said to be useful to stimulate the heart and to treat angina pectoris. The product never caught on and was replaced by more effective medications, but mescaline, along with hallucinogenic alkaloids such as psilocin and lysergic acid diethylamide (LSD), has been used in experimental psychiatry.

Although Native Americans were using peyote in sacred rituals long before any drug laws were enacted, the U.S. Supreme Court held in 1990 that such use of peyote was not constitutionally protected. Two Native Americans had been dismissed from state employment because of their use of a "controlled substance." The Indians argued that peyote is a sacred plant and that their use of it

A Huichol Indian yarn painting of the sacred peyote cactus. Note the richness of the colors that appear in this painting, which offers a sense of the visions seen during the use of peyote.

in religious rituals was protected by the "free exercise" clause of the First Amendment. The Supreme Court ruled that the religious use of peyote was irrelevant because the state law was "not an attempt to regulate religious beliefs." The Yale law professor Stephen Carter notes,

> One can understand the Court's worry about how to stay off of the slippery slope—if peyote, why not cocaine? If the Native American Church, why not the Matchbook Cover Church of the Holy Peyote Plant?—but the implications of the decision are unsettling. If the state bears no special burden to justify its infringement on religious practice, as long as the challenged statute is a neutral one, the only protection a religious group receives is against legislation directed at that group. . . . The judgment against the Native American Church, however, demonstrates that the political process will protect only mainstream religions, not the many smaller groups that exist at the margins.

Should the government ban the use of peyote by Native Americans? Many mainstream religions—Baptist, Catholic, Episcopalian, Jewish, Mormon, Methodist, and Muslim—as well as the American Civil Liberties Union and a variety of other secular organizations were concerned enough about this intrusion of government into religious affairs to file *amicus curiae* briefs in support of the Native Americans. After the case had been lost in the Supreme Court, the churches vigorously lobbied Congress to pass the Religious Freedom Restoration Act (RFRA). This law, passed in 1993 by overwhelming majorities in both the House and the Senate, requires the government to demonstrate a compelling need to justify intrusion into any religious practice.

In passing the RFRA, Congress has effectively carved out a space for the sacred in American life, a space that is in fact large enough to allow members of the Native American Church to receive the peyote sacrament without interference. However, the concerns of Native Americans and other indigenous groups embrace more than a desire to protect their sacred psychoactive plants. In the case of *Lyng* v. *Northwest Indian Protective Cemetery Association,* a Native American tribe attempted to stop the Forest Service from building a road through an area long used for sacred rituals. Though this case was tried—and lost—before the passage of the RFRA, the very nature of the case illustrates a fundamental underpinning of indigenous beliefs: the sacredness of life extends beyond that of an individual and into the realm of entire ecosystems.

To stop logging, Thai priests have wrapped the saffron robe around rain forest trees and ordained them to be Buddhist monks. In the American Southwest, the Navajo nation has protested the construction of power lines through their sacred mountains. And in Africa, tribal elders have sought to protect sacred groves, or Kayas, against resort development. Often political leaders have openly doubted that the motives for such expressions are religious rather than obstructionist. But ethnobotanists throughout the world have been struck by a single theme shared by nearly every indigenous group they have encountered: indigenous peoples believe that the entire planet is sacred.

By maintaining living collections of a diverse range of plant species, botanical gardens play an important role in the preservation of biological diversity. Here, a curator of a medicinal plant garden in Kerala, India, is caring for *Rauvolfia serpentina,* a valuable plant used in both modern and traditional medicine.

Biological Conservation and Ethnobotany

Although small by Amazonian standards (approximately 5000 hectares), the Tafua forest on the island of Savaii, Western Samoa, is precious because of the unique diversity of its life forms. Over 25 percent of the forest plants are found nowhere else on earth. Within a short radius of Mount Tofua, the small volcanic cinder cone on the eastern tip of the Tafua Peninsula, are found 132 species of indigenous plants; nine species of lizards and skinks; two species of flying foxes, one of them the endangered Samoan flying fox, *Pteropus samoensis* [Pteropididae]; and 24 species of birds. In the quiescent volcano's crater live a breeding pair of Samoan tooth-billed pigeons, *Didunculus strigirostris* [Columbidae].

The marine environments surrounding the Tafua Peninsula are also unique. Two species of sea turtles are common to the area. Dolphins, whales, and a

Tiny Tafua village in Savaii Island controls a peninsula that contains remarkable coral reefs, rain forests, flying foxes, and rare tooth-billed pigeons. The paramount chief and orator of Tafua, Ulu Taufa'asisina, has refused numerous offers of money from logging companies for the precious rain forest, condemning his village and family to remorseless poverty.

stunning diversity of reef fish and corals live within a stone's throw of the rain forest, which extends to the edge of jagged black volcanic cliffs that plunge into the sea. Interspersed among the cliffs are small coves with white sand beaches. The largest cove is the site of Tafua village.

For hundreds of years Savaii Island, like the rest of Samoa, was, in Somerset Maugham's words, "lovely, lost, and half a world away." Increasing prices of rain forest timber in the early 1970s ended that isolation. An American timber firm built a large sawmill in Asau, Savaii. Lack of experience with both the forest and the culture of Samoa eventually caused the firm to jettison the project. The sawmill, however, remained in operation under various owners, and the rain forests of Savaii continued to disappear. Now more than 80 percent of Savaii's forests are gone forever; only two large tracts of lowland rain forest remain, one of them the Tafua rain forest. Because of its proximity to the wharf, the Tafua forest offered the most lucrative logging opportunity in all of Samoa. The logging companies faced only one small problem: the paramount orator and chief, Ulu Taufa'asisina.

Samoans are gentle but determined people; but even by Samoan standards, Ulu was resolute. Although Tafua is a very poor village, with no running water, electricity, or graded road and few sources of income, Ulu steadfastly refused all offers of cash from the loggers. Not a single tree could be cut. The villagers begged Ulu to accept the logging companies' generous offers. How else could the village pay for a decent school for their children or a clinic for their sick and elderly? The loggers might even hire some of the villagers to work for them. Ulu's stance mystified the logging companies' representatives too. They were offering the village what would probably be its only chance for economic development. The loggers failed to realize that no inducement could ever persuade Ulu Taufa'asisina to allow logging. When his father lay dying, Ulu had promised to honor his last wish: Ulu had pledged to protect the rain forest with his life.

Ulu Taufa'asisina has paid a price for conservation that few individuals in industrialized countries can comprehend: he has knowingly condemned his family, friends, and village to poverty rather than accept money from loggers. "Five times the logging companies have been here asking for our forest," Ulu explains.

> I was deeply depressed because they put a lot of pressure on all of us, persuading the people of my village to sell the forest for a few dollars. I resisted, because I love my people and the land more than the money.
>
> The land is our lives. The land is also our mother. The land is sacred. I believe that the land has provided the culture, the food, the water, and the other things essential for my people. I deeply respect the honor that has been given to me, as chief orator, to become a caretaker of our beloved land.
>
> My forefathers had a dream. They had a dream that one day the land and the rain forest would be saved for eternity. They had a dream that the land and the sea would forever be well looked after, and not destroyed and distributed to other people. I share that dream. I believe that we can become masters of our destiny if we take care of our environment.

That destiny encompasses tooth-billed pigeons, flying foxes, and dolphins—and relentless economic hardship. But Ulu Taufa'asisina, like many other indigenous leaders, sees conservation in terms that transcend economic or political issues.

Indigenous Perspectives on Conservation

In many European myths, the primeval forest is the abode of witches and dragons, and is to be avoided at all costs. Indeed, the English word "savage" comes from the Latin *sylvaticus,* "of the woods." Perhaps because the forest was so greatly feared, the value of conserving it was not appreciated until relatively

Ulu Taufa' asisina, a gentle but determined man, promised to protect the Tafua rainforest with his life. He recently received the Indigenous Conservationist of the Year award from the Seacology Foundation.

recent times. Despite a high rate of deforestation in England, it was not until the reign of Charles I, in the seventeenth century, that the crown attempted to institute conservation measures. In the Western tradition, natural resources are property and therefore subject to either private or government ownership. Thus Western conservation has its roots in the pragmatic use of property; according to this viewpoint, no action should be taken that decreases the value of the resource for the long term.

Many indigenous cultures, in contrast, perceive the earth as existing not in the realm of the profane, but in the realm of the sacred, a worldview that distinguishes them from many Western traditions. Indigenous legends emphasize the need to protect the earth not because it is useful to humans but because it is sacred. The perception of conservation as a religious duty, of course, also serves ecological and cultural purposes.

The insular systems in which Polynesian cultures developed imposed harsh and rigorous penalties for environmental degradation. Although Polynesian societies initially caused the extinction of some island species (such as the moa birds of New Zealand), in time many of those societies developed cultural proscriptions against the overuse and destruction of resources. The ecological reasons for these cultural adaptations are clear. Unlike continental peoples, who could always move on, island peoples were constrained by the limits of their island homes. Degradation of resources rapidly translated into an ecosystem that could support fewer and fewer people.

Polynesian societies developed a strong ethic of conservation and wise land use. They viewed the land, including the natural plant and animal populations that occupied it, as a sacred trust inherited from their ancestors. Private ownership of land did not develop; instead, communal land tenure systems evolved. The forest and sea were not considered personal property; instead the Polynesians viewed themselves as stewards rather than owners. Chiefs acted as resource managers, accountable for their decisions not only to their contemporaries but also to their dead ancestors and to future generations. The religious system called *tapu* was used to protect resources considered particularly vulnerable. Polynesians would rather die than break *tapu,* so any resource protected by *tapu* was considered inviolate. The remnants of this land tenure system can be seen in the South Pacific today. In Samoa, Tonga, and Fiji, communal lands can be neither bought nor sold; putting a monetary value on land is perceived as incompatible with its sacredness.

Though they start from very different assumptions, both Western conservationists and many indigenous peoples recognize the need to protect vanishing natural habitats. When Maori elders became concerned about the loss of native plants used in weaving, for example, they organized a *hui,* or traditional conference, with the New Zealand Division of Scientific and Industrial Research (DSIR). They invited both scientists and traditional leaders to discuss conservation strategies. Such collaborations, although complicated by cultural differences, have provided strong support for three positions advocated by indigenous peoples: that all forest plants have a purpose and value; that the true economic (let alone cultural and spiritual) values of rain forests and native habitats have scarcely been considered and are vastly underestimated; and that entire cultures and ways of life will disappear if rain forests are destroyed. Recent ethnobotanical studies provide evidence that substantiates these indigenous views.

Quantitative Ethnobotany in South America

Forest-dwelling peoples often claim that most, and perhaps all, plants in their environment have a use. Ethnobotanist Brian Boom of The New York Botanical Garden used some pioneering plant census techniques to test that hypothesis. Working for an extended period in the Bolivian Amazon, Boom found that the Chácabo Indians knew of 360 species of vascular plants in the forest surrounding their village of Alto Ivón and that they had uses for 305 of them. They collected Brazil nuts (*Bertholletia excelsa* [Lecythidaceae]) for their own consumption and for sale, for instance, and used a plant they called *maichaca* (*Anthurium gracile* [Araceae]) to cure appendicitis. Boom then surveyed a 1-hectare plot in the tropical forest and found that 82 percent of the tree species growing there had uses known to the Chácabo. When he measured the densities of plants in the plot, Boom found that the Chácabo used 95 percent of the individual trees for some purpose.

Similar studies were undertaken by William Balée among the Ka'apor and Tembé Indians in Brazil and by Boom among the Panare Indians in Venezuela. The percentage of tree species put to use by the Ka'apor was found to be 76.8 percent, by the Tembé 61.3 percent, and by the Panare 48.6 percent. Although these findings do not prove that every forest plant has a use, they do confirm the local people's claim that the forest plants have far more uses than Western investigators have realized. Balée and Boom have concluded that certain plant families are so important in these Neotropical areas that conserving them is essential if people are to continue to depend on the forest for their sustenance.

Among these valuable plant families are the palm family [Arecaceae], the Brazil nut family [Lecythidaceae], a tropical relative of the rose family [Chrysobalanaceae], and the family that includes the hallucinogenic *caapi* vine, Malpighiaceae. These studies were the first to use a quantitative approach to demonstrate the value of the forest to indigenous people and thus to promote conservation by making clear its utility.

Working in Tambopata, Peru, with mestizo people, Oliver Phillips and the late Alwyn Gentry of the Missouri Botanical Garden employed an even more detailed quantitative technique to inventory the plants in plots in seven different types of forest. They calculated the importance of plant families used for construction, in commerce, for food, for technology, and for medicine. Their interviews with 29 field guides yielded a total of 1885 reported uses for the 605 tagged plants in their plots. When they compared data to determine whether the age of the person interviewed affected his or her knowledge of plant use, they found that the bulk of the information in some categories, such as medicinal plant lore, was held by older people. These are the people, Phillips and Gentry concluded, who should be the main focus of ethnobotanical studies and conservation efforts. Through their use of statistical tools, they substantiated the intuitive judgments of many other workers, who perceived that the long chain of oral ethnomedical tradition was coming unraveled in the current generation. Once investigators can identify the best sources of ethnobotanical information in a community or indigenous society, both local people and ethnoscientists can make more efficient efforts to conserve such information.

Forests Are More than Timber: Ethnobotanical Valuation Studies

Historically rain forests have been cut down because the simplest and quickest way to convert them into cash is to harvest the timber, burn all that remains, and plant an annual crop for a few seasons, until much of the soil's nutrients are leached out. Since most of the nutrients in the tropical rain forest are found in plant and animal tissue rather than in the soil, large-scale removal of that living material (called biomass) prevents rain forests from ever growing back. By using the tools of economics to analyze the value of land under various uses, ethnobotanists have found that in some areas there are viable alternatives to clear-cutting.

The first such study to value the nontimber resources of a hectare of rain forest was carried out in the late 1980s by Charles Peters, Alwyn Gentry, and

Each year vast quantities of tropical hardwoods such as mahogany (*Swietenia macrophylla* [Meliaceae]) are harvested from neotropical forests, but in most areas no attempt is made to utilize sustainable harvest techniques.

Annual Yield and Market Value (U.S. dollars) of Fruit and Latex Produced on 1 Hectare of Forest at Mishana, Río Nanay, Peru

Common Name	Species	Family	Number of Trees	Production per Tree	Unit Price	Total Market Value
Aguaje	*Mauritia flexuosa*	Arecaceae	8	88.8 kg	$10.00/40 kg	$177.60
Aguajillo	*Mauritiella peruviana*	Arecaceae	25	30.0 kg	4.00/40 kg	75.00
Charichuelo	*Rheedia* spp.	Clusiaceae	2	100 fruits	0.15/20 fruits	1.50
Leche huayo	*Couma macrocarpa*	Apocynaceae	2	1,060 fruits	0.10/3 fruits	70.67
Masaranduba	*Manilkara guianensis*	Sapotaceae	1	500 fruits	0.15/20 fruits	3.75
Naranjo podrido	*Parahancornea peruviana*	Apocynaceae	3	150 fruits	0.25/fruit	112.50
Sacha cacao	*Theobroma subincana*	Sterculiaceae	3	50 fruits	0.15/fruit	22.50
Shimbillo	*Inga* spp.	Mimosaceae	9	200 fruits	1.50/100 fruits	27.00
Shiringa	*Hevea guianensis*	Euphorbiaceae	24	2.0 kg	1.20/kg	57.60
Sinamillo	*Oenocarpus mapora*	Arecaceae	1	3000 fruits	0.15/20 fruits	22.50
Tamamuri	*Brosimum rubescens*	Moraceae	3	500 fruits	0.15/20 fruits	11.25
Ungurahui	*Jessenia bataua*	Arecaceae	36	36.8 kg	3.50/40 kg	115.92
Total			117			697.79

Note: The yields for *M. flexuosa, J. bataua, P. peruviana*, and *C. macrocarpa* were measured. The yields for other fruit trees are estimated on the basis of interviews with local collectors.

Robert Mendelsohn, an interdisciplinary team comprising an ecologist, a botanist, and an economist, respectively. Concerned that only sawlogs and pulpwood resources were being valued in appraisals of the worth of tropical forests, they initiated a study that also valued such products as edible fruits, oils, latex, and fiber. Working with a 1-hectare plot of lowland forest in the Peruvian

Amazon, they calculated the annual yield and market value of the fruit and latex produced there (shown in the table on the previous page). By using a simple economic model they estimated the net revenue (the profit that remains after all related costs have been deducted) that would be generated by all future harvesting of those fruits and latex. They also estimated that 25 percent of the fruit crop would remain in the forest each year to provide seedlings that would grow into fruit- and latex-bearing trees. Using these assumptions, they calculated a net present value (value of the sum of future income in today's dollars) of $6330 for the fruit and latex yielded by a hectare of forest. They calculated sustainable timber harvests to have a net present value of $490 per hectare, for a total net present value of $6820.

Comparable analysis of tree farming in Brazil gave yields of $3184 per hectare, while the conversion of tropical forest for cattle pasture gave a net present value of only $2960, assuming gross revenues of $148 per year. The researchers concluded that nonwood forest products "yield higher net revenues per hectare than timber, but they can also be harvested with considerably less damage to the ecosystem. Without question, the sustainable exploitation of non-wood forest resources represents the most immediate and profitable method for integrating the use and conservation of Amazonian forests."

In a similar study, Balick and Mendelsohn valued the native medicinal plant species taken by the local people from a forest in Belize. From two separate 1-hectare plots of 30- and 50-year-old forest, respectively, total biomass of 308.6 and 1433.6 kilograms (dry weight) of medicines was collected. It was suggested that harvesting the medicinal plants from a hectare of forest would yield the collector $564 and $3054 in the local markets, respectively, for the two plots, after the costs of harvesting, processing, and shipping were subtracted. For the 30-year rotation, a present value of $726 per hectare was calculated for the medicine, and for the 50-year rotation a present value of $3327 per hectare was calculated.

This study fostered a greater understanding of the value of the tropical forest to the local inhabitants and their economy. It ultimately led to the development of several industries based on the extraction of medicinal plants from the forest for processing into tinctures, extracts, and salves. Today local Belizean brands of traditional medicines—Agapi, Rainforest Remedies, Rainforest Rescue, Triple Moon—all help to generate employment for many local people.

Additional studies to establish the net present values of forests in other parts of the Neotropics have confirmed the relatively high value (often several thousands of dollars) of the products that can be harvested on a hectare in areas where land is now priced in the hundreds of dollars or less. Critics of this

Cacique Romão, leader of an Apinaye Indian village in Brazil, demonstrates the harvest of jaborandi (*Pilocarpus microphyllus* [Rutaceae]). Indigenous peoples have long harvested this plant from forests in northern Brazil for the extraction of pilocarpine hydrochloride, an important therapy for glaucoma and dry mouth syndrome. The plant has been domesticated in the last few years and introduced into plantation cultivation; a few hundred hectares of the crop can now meet commercial demand for the species, reducing the market for wild-harvested leaves.

method of establishing the value of forests point out that the land must be near to a market or to a distribution channel in order for the economic benefit to be realized. They state that there is probably a finite market for the commodities produced under these management schemes. Both points have their validity, but nevertheless, it is clear that in the areas that have been intensively studied the harvesting of nontimber forest products has increased the income levels of local people and has stimulated the development of new industries with a local value-added component that increases returns to the region or country of origin. Such studies also tend to confirm indigenous beliefs that tropical forests, if properly managed, have far more value than as mere sources of lumber and wood pulp. Given the proven profitability of sustainable exploitation of nontimber forest products, why has so little been done to promote the marketing, processing, and development of these valuable resources? We believe that the problem lies not in the actual value of these resources, but in the failure of public policy to recognize it.

Chicle, collected from *Manilkara zapota* trees in Central America and Mexico, is an example of an important nontimber forest product. After harvest, the fresh latex is heated for processing into coagulated blocks, and then used for various products, including a base for chewing gum.

"Green" industries now promote the sale of rain forest products, such as the buttons fashioned from palm seeds that adorn garments made from Paris to Hong Kong, ice creams flavored with exotic nuts and fruits, and rare tropical essences in perfumes, shampoos, and body creams. We can wash with soap made from tropical oils and nectars, eat cereals based on grains that once sustained the Aztecs, and drive to work on tires manufactured from wild rubber harvested in the Amazon Basin. Like other suggested solutions to the dilemmas posed by deforestation and economic development, green marketing is hardly a panacea. The continued use of these products depends on the reliability of their supply, markets, and distribution. Key to maintaining supply is the issue of sustainable resource production.

Goods from the Woods: Sustainable Production

Of greatest concern in the development of products based on tropical forest species (usually known as nontimber forest products, or NTFPs) is our ability to ensure sustainability. But nowadays, the concept of sustainability is used in a rather cavalier fashion. In truth, we know very little about the sustainability of any production, especially of the products from tropical ecosystems. When the organic foods industry was first being developed, someone observed that more organic produce was being sold than was actually being produced; the same can be said for "sustainably produced products" from the rain forest. The ecologist Charles Peters of The New York Botanical Garden has undertaken many detailed studies of tropical forest trees in efforts to determine the level of sustainable production or harvest of each species. According to Peters, "a sustainable system for exploiting non-timber forest resources is one in which fruits, nuts, latexes, and other products can be harvested indefinitely from a limited area of forest with negligible impact on the structure and dynamics of the plant populations being exploited." A plant such as *Brosimum alicastrum* [Moraceae], a tree found in Central and South America that is exploited for its protein-rich fruits, needs to produce over 1.5 million seeds to ensure that *one* tree will live long enough to reproduce. If most of the fruits produced by this species were to be harvested rather than left to grow in the forest, the population would become extinct within one generation.

Too little is known about the levels of sustainable harvest of many of the internationally important NTFPs, including the Brazil nut. Some 200,000 people harvest the Brazil nut from the 20 million hectares of Amazon forest where it grows; annually they produce around 42,000 metric tons for the commercial trade, valued at approximately $35 million, or 1.5 percent of the total international nut trade. The harvest of this nut is one of the largest sources of cash income for many of these people, and reduction in government subsidies for other NTFPs such as rubber, which grows in the same areas, has led people to harvest increased quantities of Brazil nuts. What, then, will happen 50 or 100 years from now, when most of the seeds produced by once-great populations of Brazil nut trees have been removed from the forest and sold? Quite simply, the mature, seed-producing trees that are the backbone of the population will die

Palms often have multiple uses. The palm açaí (*Euterpe oleraceae*), for example, has numerous uses, for both subsistence and commercial purposes, that derive from different parts of the palm.

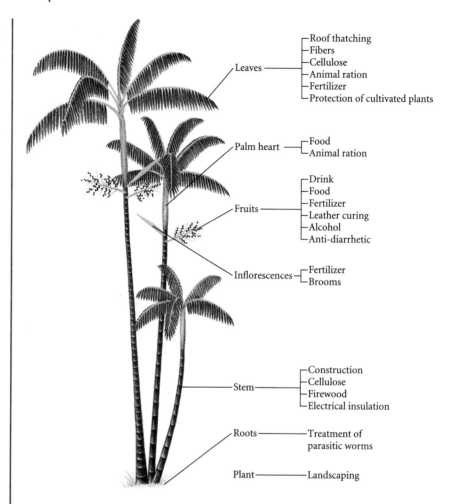

Leaves — Roof thatching / Fibers / Cellulose / Animal ration / Fertilizer / Protection of cultivated plants

Palm heart — Food / Animal ration

Fruits — Drink / Food / Fertilizer / Leather curing / Alcohol / Anti-diarrhetic

Inflorescences — Fertilizer / Brooms

Stem — Construction / Cellulose / Firewood / Electrical insulation

Roots — Treatment of parasitic worms

Plant — Landscaping

and not be replaced, and the resource base on which these industries are built will disappear.

In many parts of the world, NTFPs are an important component of the local economy. The heart of palm, for example, sold as a gourmet food, is often harvested from a tree native to the Amazon estuary of Brazil, where it is known as *açaí* (*Euterpe oleracea* [Arecaceae]). In that region, the harvest, packaging, and sale of palm hearts employ nearly 30,000 people and generate $300 million each year. The palm also provides a wealth of other subsistence and communal products, as illustrated in the figure on this page. The *açaí* palm is the most useful tree species in some 25,000 square kilometers of forest that is subject to seasonal flooding. Each hectare has up to 7500 palms in various life stages. These flood-prone forests in Brazil produce around 200,000 metric tons of palm heart

per year, over 95 percent of which are consumed in Brazil. The bulk of the ex-
ports go to Europe. Most of the harvesting is carried out in a destructive fashion: the harvesters remove the maximum number of stems from these multi-
trunked trees, with little concern for their regeneration.

Ecologists such as Anthony Anderson have studied how the local people
manage the palm and have measured economic returns. One family Anderson
worked with harvested almost $3000 worth of hearts and $15,532 worth of edible fruits from their *açai* palms in a single year. The family was adept at managing the palm forest in a sustainable manner. Elsewhere harvesters are commonly paid by the piece, and they extract the product without regard to future
harvests. A harvester is said to be able to cut between 150 and 200 palms a day
to extract their hearts. A survey of current production shows that smaller and
smaller stems are being harvested, an indication that the time elapsing between
harvests is growing shorter. Thus, unless *açai* palms are replanted in the forests
or in plantations or unless harvesting strategies are modified, the trees will vanish. Tragically for the forest, for the people who live there, and for those who
consume its products, this is often the case with NTFPs. When plants harvested
for medicines disappear, patients will suffer. Benign prostatic hyperplasia, for
example, a debilitating condition of middle-aged and elderly men, can be
treated successfully with a drug manufactured from the bark of *Prunus africana*
[Rosaceae], a tree that grows in the forests of Cameroon, Zaire, Kenya, and
Madagascar. Although techniques of sustainable harvest have been developed,
they are not always employed, and this important resource, along with the
fauna it supports, is in decline.

What, then, are the options for the continued use of NTFPs as a tool for economic development and conservation of biodiversity in the future? Charles
Peters suggests six steps for exploiting NTFPs in a sustainable fashion. First, the
species to be exploited should be carefully selected, after such factors as the ease
of harvesting and resilience of natural populations to disturbance are considered. A tree valued for its roots will be harder to harvest than one valued for its
fruits, and the harvest of a species that produces fruits in massive quantities at
one time of year will be easier to manage than the harvest of a species that produces fruits sporadically throughout the year. Once the species has been decided upon, a forest inventory should be undertaken to learn where the resource is found in greatest abundance and the number of productive plants per
hectare. Investigators then should estimate the quantity of the resource produced by the species in its various habitats and by trees in all size classifications,
to determine which trees in which habitat it is best to harvest.

When these three steps have been taken, the harvesting of the resource can begin, but the careful measurement should continue. The status of the population

should be monitored for signs that the forest is being overharvested. People should examine the status of adult trees periodically to determine whether the flowers are being pollinated, whether large numbers of fruits are being consumed by predators, and so on. If problems arise, the harvest should be adjusted to keep its level below the rate that would threaten sustainability.

When necessary, people may replant areas that do not seem to be regenerating, clean out competitive species, or open up the forest canopies to allow more light to reach the young trees and thus speed their regrowth. The precise measurements that Peters recommends are expensive and time-consuming, and very few species have been studied from this perspective. However, plant populations may be threatened if harvests are determined by the demands of the marketplace rather than the needs of the ecosystem. As Peters notes, "nature does not offer a free lunch." In our enthusiasm to support conservation of the natural world by focusing on its usefulness to economies, we are perhaps inadvertently dooming elements of it to extinction. Only when ecologically sound management plans based on scientific studies are developed for resource extraction will the use of those resources be able to contribute to the conservation of biological diversity.

Cultural Ties to the Forest: Palms, People, and Progress

Worldwide, more than 200 genera of palms (the family Arecaceae) have been described. Wherever they grow, in desert, in humid forest, or on cool mountain slopes, palms make important contributions to human sustenance and culture. Because of their ability to colonize degraded or damaged tropical ecosystems, palms can assume singular importance as a source of food, building material, and income. Many traditional peoples view the coconut palm as a gift of the gods; palms may prove to be another kind of gift in their ability to ameliorate the consequences of environmental degradation to tropical peoples.

Because of their ability to adapt to sites that are inhospitable to other species and their usefulness to local people for the vast assortment of commercial and subsistence products they provide, palms are increasingly becoming a prime focus of resource management studies in tropical ecosystems. "The versatility of palms in the hands of man is astonishing," noted the late Harold E. Moore, Jr., who was the greatest authority on palms in the twentieth century.

> Houses, baskets, mats, hammocks, cradles, quivers, packbaskets, impromptu shelters, blowpipes, bows, starch, medicines, magic, perfume—all are derived from palms. . . . To whatever extent man

has been involved in the tropical ecosystem, palms have certainly been a major factor in making possible this involvement and even today, despite the advent of the corrugated tin roof and the rifle, they are of primary importance to many primitive American cultures.

"Oligarchic" tropical forests—those that are dominated by only a few plant species—characterize many tropical landscapes, particularly where a swampy or degraded or seasonally flooded habitat makes it difficult for a diversity of forest species to grow. Species of palms that local people can exploit often predominate in the oligarchic forests of the lowland tropics. Local people from Australia to Southeast Asia sell the edible sap of the nipa palm (*Nypa fruticans*), found in swamps along the coasts. Freshwater swamps on Pacific islands and in Southeast Asia are home to the sago palm (*Metroxylon sagu*), from which local people extract edible starch. In the Neotropics an important starch-bearing palm is "moriche" (*Mauritia flexuosa*), which forms stands in freshwater swamps in northern South America.

The babassu palm (*Orbignya phalerata*) is found in low densities in lowland Neotropical forests such as those in Brazil, where its fruits are harvested as a source of edible oil and charcoal for cooking. When these forests are cut down

The nipa palm, commonly found along riverbanks and in swamps, is the source of a palm sap consumed as a beverage. The leaves are also a good source of thatch, said to be more durable than coconut leaves.

Left: The fruits of the babassu palm (*Orbignya phalerata*) are harvested from wild groves throughout the palm's extensive distribution in the northeast of Brazil and elsewhere. Fruits, weighing up to 250 grams or more each, are gathered in baskets or sacks and transported to the collectors' homes for processing. Right: Babassu fruits are cracked on the head of an upturned ax, and the oil-rich kernels are extracted from the woody center, or endocarp. These kernels are then sold to a factory for processing into edible oil.

for agriculture, however, the babassu palm quickly comes to dominate the landscape. People manage these oligarchic stands of babassu for the harvest of their fruits and for dozens of other subsistence products. The degraded soils of northeastern Brazil, many of which are abandoned agricultural lands, are covered with vast stands of babassu palms, which support the largest oilseed industry in the world that is completely dependent on a wild plant resource. The harvest and processing of these oil-rich fruits involves more than 1 million people. Very few other economically important plant species can tolerate the searing sun, prolonged dry season, and intense flooding from torrential rains that characterize northeastern Brazil.

Lengthy field research is necessary to appreciate and understand fully the power and durability of the bond between palms and people. A number of studies have shown how indigenous peoples' management of palm resources can serve to support the conservation of their ecosystem.

The palm genus *Sabal* is widely used in the Yucatan Peninsula of Mexico and in nearby areas. The descendants of the Maya manage the palms as a source of construction material, firewood, food, medicine, and magic. When ethnobotanist Javier Caballero of the Jardín Botánico de la Universidad Nacional Autónoma de México investigated the use of *Sabal* species in this region, he found that some uses have become extinct (magic and medicine), others are declining (brooms, poles, fences), while other uses persist (thatch, fuel) or are increasing (handicrafts). He documented the great pressure on the palm resource for thatch. The 3500 to 5000 *Sabal* leaves needed to thatch a single house require the harvest of 250 to 1250 trees (ranging from juvenile to adult). Caballero's study showed the importance of applying resource management techniques to ensure future supplies of palm leaves and the dangers posed by overexploitation of the resource.

Just as shamanistic teachings or techniques of herbal healing have been refined over many generations of experimentation, so too have technologies for managing resources in environments where conventional Western methods fail. Ethnobotanical studies of the ways forest resources have traditionally been managed can offer practical alternatives for regions where unsuitable land is increasingly being employed for conventional agriculture. One conservation technique emerging from ethnobotanical studies, although somewhat controversial, involves the creation of extractive reserves and indigenous-controlled reserves.

Conservation Areas and Indigenous Peoples

Early nature preserves were established in the tropics during colonial times primarily to serve the needs of big-game hunters or to protect watershed and timber resources. The colonial administrations created most existing rain forest reserves by simply declaring government land to be a national park or by purchasing land from private owners, the same strategies followed to create national parks in North America and Europe.

The Mexican government owned more than 99 percent of the 528,000 hectares of rain forest, wetlands, and coral reefs that are now included in the Sian Ka'an Biosphere Reserve on the Yucatan Peninsula, whereas the 100,000-hectare Guanacaste National Park in Costa Rica, an area of dry lowland tropical forest, was purchased largely from private landowners for $9 million contributed by a variety of conservation organizations, trusts, foundations, private donors, and government agencies.

The Monteverde Cloud Forest Reserve in Costa Rica, on the other hand, was created partially through a "debt-for-nature" swap—conservation organizations purchased part of the country's international debt and accepted conservation of rain forest acreage as payment. While these strategies have been effective

The milky latex of the rubber tree is gathered by wounding the stem and collecting the sap in a small vessel. The rubber tapper then pours the collected latex into a larger container and processes it into a solid block or ball.

in preserving land, they have been primarily focused on meeting national needs rather than the concerns of local peoples.

The government's power to appropriate land for national parks under the right of eminent domain, the method used in part to create Grand Teton National Park in the United States, may have the most severe cultural consequences for indigenous peoples. Poaching frequently becomes a problem when indigenous peoples or long-term residents are displaced from their ancestral lands.

These traditional strategies create reserves that are essentially free from human disturbance. New strategies differ in that they emphasize the possibility of using the resource while protecting it from degradation. In the late 1980s, Brazil created a category of forest reserve known as the "extractive reserve," an area where local people can extract products on a small scale while still preserving a largely intact ecosystem. This form of biological reserve is closely associated with a social movement, begun in the state of Acre, which attempts to improve economic standards among Brazil's traditional peoples. The first reserves were established for the extraction of rubber and Brazil nuts. Most of the rubber produced in the Amazon Basin is gathered in a way that does not destroy the trees, so the people who gather it are strongly opposed to any destruction of the rain forest. When ranchers who wished to clear the forest assassinated Chico Mendes, the movement's most visible leader and organizer of the local rubber tappers' union, there was such an outcry that the government eventually responded by creating the first major extractive reserves, which now total some 10 percent of the entire state of Acre. Over the last few years, however, the value of both wild-harvested rubber and Brazil nuts has fallen. Ethnobotanists and taxonomists such as Douglas Daly of the New York Botanical Garden are working with local inhabitants of the reserves to identify other species that can be produced for regional and international commerce and provide income opportunities for the people who protect these areas of tropical forest. Individuals in north temperate countries are also helping to create extractive reserves. Such organizations as Conservation International and Cultural Survival have organized the marketing of forest products from reserves.

Maintaining an extractive reserve is not without its problems. The forest ecosystem may be damaged if economically important products are overharvested. As we have seen in the case of the *açai* palms harvested on one family's land in Brazil, there is sufficient incentive to use sustainable harvest techniques on private lands, but what about the land owned by a community? Even if the resources are designated as communal property, will individuals place personal need over that of the community and engage in overharvesting? An extractive reserve differs from a parcel of land that is simply public property in that a social structure is a key element of the reserve. Ideally, guidelines can be developed and rules and regulations enforced. In one reserve in San Rafael, Loreto, Peru, a regional farmers' union helped villagers establish a protected area and develop rules for the communal use of the reserve, although the land was still under the control of the Ministry of Agriculture. While such rules do not exist or are not strongly enforced elsewhere, it appears that most reserves are respected by the local people, particularly if they are established with an understanding of local culture and needs.

Sometimes the ecosystem of a forest reserve can be altered by protecting economically important species (or even augmenting them by strategic replanting) while other species—those whose value is not clear—are not protected with the same tenacity. Thus, some critics have argued, extractive reserves are able to protect only a portion of the biodiversity they contain. The lesson is that there is no one formula for protecting wildlands, especially in remote regions of the tropics. But in the global effort to assemble a jigsaw puzzle of conservation areas, ethnobotanical research can play an important role by helping to preserve and disseminate traditional knowledge. When this knowledge is applied, economic returns can accrue to those who make their living in the rain forest while still protecting it.

Conflicts Between Indigenous Peoples and Nature Preserves

Because indigenous peoples have seldom been involved in the planning process, conflicts have often arisen between them and Western-educated preserve managers. Conflict has been particularly acute over the management of Amboseli National Park in Kenya, which is overgrazed and thus capable of supporting far smaller animal populations than in the past. "The decline of Amboseli has little to do with the large number of tourists that used to visit it or with the increase in elephant population—excuses often used by the authorities, for which there is ample evidence to refute," writes David Lovett Smith, a former warden of Amboseli.

> The demise has been brought about, in my opinion, by inept management and a total lack of communication with the local people. . . . For it was the Masai people who themselves looked after the wildlife until governments and wildlife authorities took over its management, and, from the 1970s on, proceeded to mismanage it so badly.

"Rules Governing the Extraction of Forest Products from the Communal Reserve in San Rafael, Loreto, Peru"

1. It is prohibited for individuals, families, and groups to extract timber from the community reserve. The extraction of timber is only permitted at the communal level. Timber extraction is permitted when the community needs money for communal purposes, such as a new school or medicine for the village.

2. Nontimber forest products can be extracted by any individual, family, or group.

3. Poles and other local construction materials can be extracted by any members of the community. Extraction for commercial purposes is limited to small quantities and to certain periods of the year, such as when rice is harvested.

4. The extraction of fruits and medicinal plants is permitted to everybody. People from the community and from the neighboring communities can extract these products either for their own consumption or for the market.

5. In collecting fruits, leaves, flowers, bark, resins, roots, and branches, cutting trees is prohibited. The extraction of particularly valuable species is regulated by special rules.

6. Timber extraction by non-members of the community requires a permit issued by the community.

7. Permits to extract timber from the communal forest reserve must be solicited from the Ministry of Agriculture in the name of the *Teniente Gobernador* of the community and must be renewed every year in Iquitos.

8. It is prohibited to use the area of the community forest reserve for agriculture. However, the community recognizes the rights of its members over their old fallows for an unlimited time.

Implicit in the authorities' explanation of Amboseli's degradation is the view that the Masai and their herds are inimical to efforts to conserve the ecosystem, rather than potential parts of the solution. But changing such a view requires reconsideration of some of the fundamental tenets of Western land ownership and management.

Consider the differences between Western and indigenous notions of property ownership among the Turkana, a pastoral African people living in the Rift Valley of northwestern Kenya, about 250 kilometers northwest of Amboseli National Park. Like many indigenous peoples, the Turkana do not believe in private ownership of natural resources. Instead, the *Acacia* [Mimosaceae] trees that their goats feed on are administered as a communal trust. Village elders ration feeding privileges, chasing away offending goats with sticks. In the Western view, such communal trusts are inherently unstable. According to most Western theorists, such resources will inevitably be degraded, and the result will be what bioethicist Garrett Hardin has termed "the tragedy of the commons."

To forestall what they saw as the "inevitable" collapse of the Turkana grazing system, a team commissioned by the United Nations divided up the Turkana grazing areas into plots, which were then deeded to private individuals. The village elders' sticks were no longer required. Soon, however, all of the *Acacia* trees were completely denuded. George Monbiot argues that while the indigenous system of communal ownership of the *Acacia* trees might not have been sustainable in a Western society, it worked for the Turkana.

A new and very important branch of ethnobotany might be termed "ethnoconservation biology"—the incorporation of indigenous conservation models into wildlands management. Attempts are now being made to document indigenous conservation strategies throughout the world. Take another conservation model from Kenya, the *Kayas,* or sacred groves, of eastern Kenya. For centuries local people have used these small forests for religious ceremonies and burials. Each is controlled and managed by the village elders. Thirty *Kayas* in the Kalifi district constitute what Alison Wilson has termed "islands of biodiversity" in a sea of agriculture. "It is a curious paradox," Wilson writes, "that historically the *Kayas* have been preserved not despite human settlement but because of it."

> Although *Kayas* are small (their combined area is around 2000 hectares) and often isolated, their value in terms of biodiversity is out of proportion to their size. The elders' protection has been a fruitful legacy, for in these tiny forests are found species of trees that have disappeared or are in great danger elsewhere. As significant as

their biological value is, though, it is the *Kayas'* cultural and historical value that may ultimately be the key to their survival.

The Kenyan government recently proposed national park status for the *Kayas*. The indigenous people, however, opposed the plan, fearing that the government might then deny them access to the *Kayas*. A compromise was reached and the *Kayas* were granted status as national monuments rather than national parks, but the village elders were right to be concerned: a *Kaya* on Chale Island has been taken over by a foreign property developer. The local elders saw this as a spiritual catastrophe:

> The spirits of our ancestors have warned us of calamity should our sacred groves be destroyed. Already, drought has come to our land and strange portents such as baboons eating goats and monkeys eating eggs have been witnessed by us. We are appealing to President Moi to help us regain our land.

Some Western environmental organizations do not hesitate to impose Western solutions on indigenous cultures whose conservation ethics predate those of the Europeans by thousands of years. Though Western culture has generated numerous environmental disasters, we often ignore, discount, or erode the wisdom of cultures that have preceded ours. Even organizations that claim to champion indigenous causes would find it extraordinarily difficult to turn over their operating budgets and decision-making authority to the Turkana or the Samoans. Like colonialists of an earlier era, Westerners sometimes make decisions "for the good" of indigenous peoples, but would never dream of allowing indigenous peoples to make such decisions for them. Such "ecocolonialism" can be just as corrosive to indigenous cultures as its political antecedents.

The alternative is conservation experiments that attempt to translate ethnobotanical studies into vehicles for conservation activities. These programs have experienced both successes and setbacks, and they reveal some complexities of conservation initiatives that attempt to bridge widely disparate cultures.

An Ethno-Biomedical Forest Reserve in Belize

Belize is a small nation with a population of around 200,000. Vast tracts of forest still cover a significant portion of the country. A recent environmental profile of Belize estimated that more than 93 percent of the country could be classified as forest land, although this estimate was optimistic, for it excluded only urban areas and large-scale farming operations. In 1988 the Belize Ethnobotany Project was initiated to inventory, understand, and conserve as much ethnobotanical data as possible in a country that is undergoing rapid change, accompanied by loss of natural habitat and the erosion of existing cultures. The project is a collaborative effort between The New York Botanical Garden's Institute of Economic Botany, and the Ix Chel Tropical Research Foundation in the Cayo District with the Belize Center for Environmental Studies, the Belize Zoo and Tropical Education Center, the Belize College of Agriculture, and a host of other governmental and nongovernmental organizations in Belize. This multi-tiered effort has linked the mutual interests and activities of local healers, farmers, students, ethnobotanists, and pharmaceutical researchers to the conservation of their main source of materials and ideas: the area's forests. These forests serve as both a classroom and a source of raw materials for local health practitioners. An ongoing inventory of species and their uses focuses on the collection, documentation, and study of traditional medicines. Collection efforts in small villages and isolated forest regions have been linked to the Developmental Therapeutics Program of the National Cancer Institute in the United States, supplying it with more than 2000 bulk samples for testing in its cancer- and AIDS-screening program.

Supplementing the more familiar role of ethnobotany as a documentary science, the project seeks to renew interest in cultural knowledge and its transmission, particularly in the area of medical practices. It has focused on work with groups of elderly healers, most of whom have no apprentices and whose accumulated knowledge is in danger of being lost. The long-term interdisciplinary nature of the project has allowed in-depth work with traditional healers in efforts to understand disease concepts, healing traditions, and the uses of plants. This type of knowledge recovery has been described as "salvage ethnobotany."

The project helped local organizations convene four national traditional healers' meetings. The open forum provided by these meetings enabled healers from different cultural groups and geographic regions of Belize for the first time to exchange information about the medical uses of local and exotic plants. They discussed the importance of traditional healing, the central role of the healer as community health care provider, and the increasing difficulty of locating certain useful species.

In 1992 the Belize Association of Traditional Healers was formed and Rosita Arvigo of the Ix Chel Tropical Research Foundation was elected its president. Yet without plants, their work is impossible. As one of the healers, Hortense Robinson, said, "we can't do our work without the plants—it's like a mechanic without his tools. Just knowing what the name of the plant was won't help—you can't use the name to heal you." As part of the effort to conserve species that are important in the work of traditional healers, a 2400-hectare parcel of

Left: Rosita Arvigo, a naprapathic physician, was an apprentice to the late Don Elijio Panti, a well-known Maya Healer in Belize, for more than a decade. Here they are shown gathering herbs along the trail near his house in San Antonio village. Right: Leopoldo Romero, an accomplished herbalist and bushmaster (a person knowledgeable about the forest or "bush" as it is called in Belize), explains the medicinal properties of mahogany to Michael Balick and Rosita Arvigo, who have undertaken ethnobotanical studies with over two dozen traditional healers as part of the Belize Ethnobotany Project.

lowland tropical forest was given forestry reserve status in June 1993 at the suggestion of a government minister, Daniel Silva, who noted that Belize has a rich tradition of conservation reserves. It has reserves for jaguars, for monkeys, for butterflies, so why not for medicinal plants? The reserve was intended to provide a source of medicinal plants as well as a place to teach apprentices. Funds for surveying and demarcating the reserve were provided by the Healing Forest Conservancy and the Rex Foundation. The forest, in the Yalbak region of Belize, contains a wide diversity of fauna as well as many useful medicinal plant species. As originally conceived, this "ethno-biomedical forest reserve" would serve as a site to promote ethnobotanical and ecological research in efforts to define harvesting regimes for sustainable extraction. Toward this end, a team of scientists is carrying out ecological inventories as well as experiments designed to learn at what rates bark and roots will regenerate after harvest. Unfortunately, within a year after the reserve was established, the local government changed, and controversy arose over which group of local healers would be responsible for its operation. Various plans have been submitted to the Forestry Department and scientific experiments continue, but development of the educational and social component of the reserve's program is currently on hold. Despite the best of intentions, not all conservation efforts are immediately successful.

As habitat destruction and overharvesting are depleting the supply of medicinal plants in the forests of Belize, Rosita Arvigo and Gregory Shropshire of the Ix Chel Tropical Research Foundation have started a program to develop horticultural nurseries in collaboration with Hugh O'Brien of the Belize College of Agriculture. As part of the program, the subject of medicinal plants was introduced into the college curriculum. The major goal of the joint project is to learn to propagate many of the commercially valuable plants currently harvested from the wild. The species differ widely in their morphology and biology—some are easily reproduced weedy herbs while others are long-lived trees—so the task is complex. Local and regional businesses that depend on native plants will ultimately benefit from techniques being developed in the nurseries. The project also rescues plants threatened by development. A team from Ix Chel and the Belize College of Agriculture collects seedlings of rare or slow-growing trees from areas soon to be cleared for housing and transplants them to a "tree orphanage." Eventually they will settle the young trees in more secure areas, such as forest reserves and privately owned farmlands.

Indigenous-Controlled Reserves in Samoa

For many years Falealupo village, on the western tip of Savaii Island, resisted all approaches by companies that wanted to log the 10,000-hectare lowland rain forest. In 1988, however, the village was forced to allow logging in order to pay for a government-mandated school. Since per capita income in Falealupo is less than $100 U.S. dollars per year, logging represented the only possible source of the funds needed to build the school. After considerable debate, the villagers decided to allow logging on a "license basis": the logging company would pay for each tree removed, and it would cease logging as soon as the village had accumulated the $65,000 required for the school.

Logging began in June 1988, and the villagers wept as the bulldozers pushed over the trees. Paul Cox, with Thomas Elmqvist of Umeå University in Sweden and William Rainey and Elizabeth Pierson of the University of California at Berkeley, was doing fieldwork in Falealupo at the time. The research team shared the villagers' alarm and despair over the destruction of the forest. Cox posed a question to the village council: Could they stop the logging if the team undertook to raise enough money abroad for the school?

The villagers met for hours, fearful that the overture represented yet another attempt by foreigners to seize their forest. Eventually, however, they decided to trust the ethnobotanists and dispatched two paramount orators to stop the loggers. In the following weeks, the research team raised funds from their families, friends, and students, from two manufacturers of herbal products—Forever

Living Products in Phoenix, Arizona, and Murdock International in Springville, Utah—and from a firm that deals in antique botanical prints, James MacEwen and Associates of London. In January 1989 a representative group of donors arrived to negotiate and sign the Falealupo Covenant. The donors renounced any rights to or interest in the land and pledged to build the school in exchange for the village's promise to protect the rain forest for 50 years. The covenant allows the villagers to continue to harvest medicinal plants and timber for kava bowls, canoes, and other cultural purposes but forbids logging or any other activity that could significantly damage the forest. The covenant also allows ethnobotanists to investigate medicinal plants in the forest but mandates that 33 percent of income from any commercial discoveries be returned to the village. This provision became the basis of the National Cancer Institute and Brigham Young University's approach to distribution of income from possible commercialization of prostratin, the anti-HIV compound.

The advantages of the Falealupo Covenant are clear. Relatively small sums of money were required to secure the preserve, far less than would be needed to purchase the land. Donated funds were used to construct needed public works. This allows donors to achieve two worthy goals, namely construction of a school and preservation of the rain forest. The covenant relationship entails few enforcement or survey problems; the villagers, who by covenant are responsible for protection of the forest, have very effective enforcement and judicial proce-

The Falealupo rain forest preserve contains one of the last large tracts of lowland rain forest remaining in Samoa. The Falealupo villagers established the reserve, with the cooperation of overseas donors, in return for assistance to build a needed school. It is home to an endangered species of flying fox.

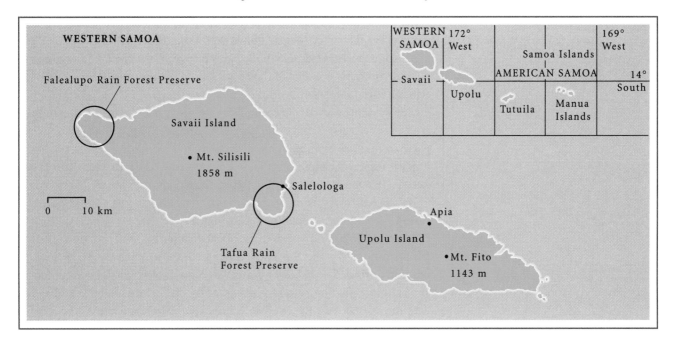

dures. Not least, the management and administration of the reserve are completely in the hands of the villagers, who are heirs to several thousands of years of management techniques. As a result, the reserve management is compatible with indigenous conservation and spiritual values.

Nevertheless, the establishment of the Falealupo Reserve was not without complexities. Establishment of the covenant required a sophisticated understanding of the language and the culture. The covenant was sealed by ritual rhetoric and consumption of kava, as Samoan pledges have always been made. The long negotiations preceding the covenant were conducted entirely in the Samoan language and according to Samoan customs, so all parties involved had to be fluent in the language. The requirement that donors renounce all rights to the land is controversial, since it conflicts with current environmental thought as well as with Western legal practice. Finally, establishment of a covenant-based reserve requires the villagers and the donors to trust one another. In Falealupo the villagers feared that the donors might seize their lands and the donors feared that the villagers might eventually renege on their promise to protect the forest.

The Falealupo Reserve encountered outside criticism as well. Some Western environmentalists criticized it because continued use of forest resources by the villagers conflicts with Western concepts of conservation. Other Westerners,

Two rain forest preserves, completely owned, controlled, and administered by villagers, have recently been established on the island of Savaii in Falealupo and Tafua. These indigenous reserves complement an existing government preserve on Upolu Island.

suspicious of all ethnobotanists as "biopirates" or exploiters of indigenous peoples, believe that any covenant should be negotiated in accordance with Western legal practices rather than indigenous dictates. Yet when the time comes to pledge significant amounts of their own resources to indigenous reserves, such critics are strangely silent.

The Falealupo Covenant became extraordinarily popular in Samoa, however, both in other villages and at high levels of the government, because of its explicit recognition of indigenous sovereignty over the forest resources and its implicit respect for indigenous knowledge systems. The Falealupo Rain Forest Reserve is remarkably successful. The villagers and the donors have scrupulously abided by all terms of the covenant, and the reserve has gained widespread publicity throughout the Pacific islands.

News of the Falealupo Covenant reached Tafua, on the other side of Savaii Island. Inspired by the Falealupo experiment, a new foundation, Seacology, based in Springville, Utah, was organized to preserve Pacific island rain forests and culture. A film on Falealupo produced by the Swedish director Bo Landin, with the assistance of the Swedish Nature Foundation and the World Wildlife Fund, helped fund-raising efforts in Sweden. Soon a school and other village improvements were built in Tafua with donated funds, and the Tafua Rain Forest Reserve was created. Ulu Taufa'asisina had achieved his dream of protecting his forest forever.

The Future of Ethnobotanical Conservation

Many challenges face ethnobotanists in future years, particularly the rapid loss of biodiversity and the concomitant loss of indigenous knowledge systems. The International Union for the Conservation of Nature, for example, has placed 22,137 species of seed plants on their "Red List," indicating that about 9 percent of the world's flora is threatened with extinction. Similar data from linguists suggest that 2,400—44 percent—of the world's approximately 5,400 extant spoken languages are threatened with extinction. UCLA anthropologist Johannes Wilbert tells us that many years ago the Warao of Venezuela were highly amused when he carefully documented their traditional dances. Why, they wondered, had this man come so far to study something everyone knew how to do? When their grandchildren attempted the same dances three decades later, they turned to Wilbert to find out whether they were doing them correctly. A similar story is told by Maurice Zigmond, who studied the language of the Kawaiisu people in California in the 1940s. In the 1970s the grandchildren of the people he had worked with wrote to Zigmond to ask him to come back and

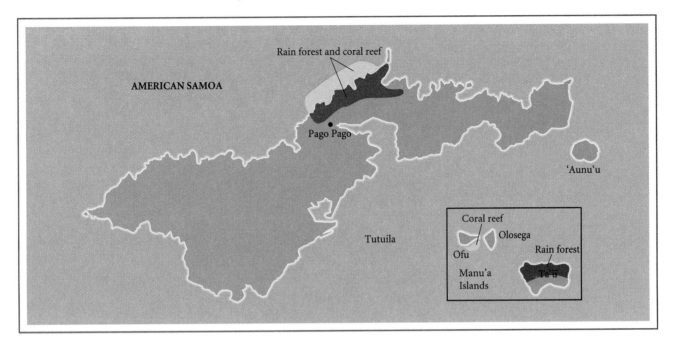

help them determine the proper pronunciation and grammar of their native language. Thus ethnobotanists preserve traditions that otherwise would surely be lost.

Some contemporary critics fear that outsiders' studies of traditional knowledge are not without risk. Published reports of the use of a medicinal plant might create a demand for the resource, with riches flowing to all parties involved except the original owners of the knowledge. Others fear that if only the most sensational information is written down, the more mundane information—about food plants, for instance—may be lost. Clearly, one priority for the future is to involve indigenous colleagues in ethnobotanical research as co-investigators and to train a new generation of people from a variety of cultures to initiate studies among their own people.

Although some colleagues scoff at the idea, the local people involved in ethnobotanical research, especially the healers, are increasingly being credited as co-authors of scientific papers and receiving patent rights to discoveries that result from the information they have provided. In ethnobotanical research, anything less than equal treatment should be viewed as unacceptable.

Given the gulf separating the ways of indigenous cultures and modern societies, it would be easy to be pessimistic that both indigenous cultures and indigenous habitats can be conserved. The tremendous economic and political

An act of Congress established the National Park of American Samoa in 1989. The park lands, consisting of rain forest and neighboring coral reef in the islands of Tutuila, Ta'ü, and Ofu, were acquired from the villagers by long-term leases, signed in 1993. Territorial Governor A. P. Lutali was named Indigenous Conservationist of the Year by the Seacology Foundation for his leadership in negotiating park legislation that was acceptable both to village chiefs and to U.S. Congressional leaders.

asymmetries between indigenous peoples and Western organizations make it almost certain that indigenous concerns will be secondary to conservation efforts. But both of us remain optimistic that solutions can be found that are acceptable to all parties. We were heartened when Congress passed an act establishing the United States' fiftieth national park in the territory of American Samoa—a park in which the government does not own a single square centimeter of ground. Samoan chiefs hesitated to relinquish communal rights to forests that they considered sacred, while Congress was reluctant to make any investment in trails, information centers, and other visitor facilities unless the government had ownership of the land. Yet a team led on the Samoan side by Territorial Governor A. P. Lutali and on the Washington side by Congressman Bruce Vento from Minnesota was able to strike a remarkable compromise: the government would acquire the rain forests and associated coral reefs by long-term leases. The legislation preserves Samoan land rights, including the right to harvest medicinal plants and other resources from the forests using traditional techniques and tools, yet gives the government right of access for 55 years— long enough to outlast the predicted life of any structure built. At the end of the lease, both parties have the option to renew.

Is there, then, a place for indigenous cultures in the twenty-first century? While we have no wish to deny modern technology to indigenous peoples, we also have no desire to see them plunged needlessly into the problems of modernity. In this, as in so many issues, we rely on indigenous wisdom. We believe that indigenous peoples, if given proper information and granted status as equal partners, are capable of plotting their own future. And while that future will likely include satellite ground stations, kidney dialysis machines, and personal computers, we are determined that the information flow should not be one-way, from Western nations to indigenous peoples. One of the most important lessons that we have learned as ethnobotanists is that plants have deeply influenced the human condition. It is our fondest hope that the richness of indigenous plant uses, and the dignity of indigenous knowledge systems, will not only continue to be part of the cultures in which they developed but will also increasingly grace our own.

Suggested Reading

General

Alexiades, M. N. 1996. *Selected Guidelines for Ethnobotanical Research: A Field Manual.* Bronx, NY: The New York Botanical Garden.

Baker, H. G. 1970. *Plants and Civilization,* 2nd edition. Belmont, CA: Wadsworth Publishing Company.

Boom, B. M. 1987. Ethnobotany of the Chácobo Indians, Beni, Bolivia. *Advances in Economic Botany*, Volume 4.

Cotton, C. M. 1996. *Ethnobotany, Principles and Applications.* Chichester, England: John Wiley & Sons.

Davis, W. 1996. *One River: Explorations and Discoveries in the Amazon Rain Forest.* New York: Simon & Schuster.

Ford, R. I. 1978. Ethnobotany: Historical diversity and synthesis. In R. I. Ford, M. F. Brown, M. Hodge, and W. L. Merrill, eds. *The Nature and Status of Ethnobotany,* pp. 33–49, No. 67. Ann Arbor: Museum of Anthropology, University of Michigan.

Heiser, C. B., Jr. 1985. *Of Plants and People.* Norman, OK: University of Oklahoma Press.

Hill, A. F. 1952. *Economic Botany: A Textbook of Useful Plants and Plant Products,* 2nd edition. New York: McGraw-Hill.

Kreig, M. G. 1964. *Green Medicine.* New York: Rand McNally.

Langenheim, J. H., and K. V. Thimann. 1982. *Botany: Plant Biology and Its Relation to Human Affairs.* New York: John Wiley & Sons.

Lewington, A. 1990. *Plants for People.* London: The Natural History Museum.

Martin, G. J. 1995. *Ethnobotany: A Methods Manual.* London: Chapman & Hall.

Simpson, B. B., and M. Conner-Ogorzaly. 1995. *Economic Botany: Plants in Our World,* 2nd edition. New York: McGraw-Hill.

Tippo, O., and W. L. Stern. 1977. *Humanistic Botany.* New York: W. W. Norton.

Chapter I: People and Plants

Anon. 1954. *The Rauwolfia Story.* Summit, NJ: Ciba Pharmaceutical Products, Inc.

Arber, A. 1938. *Herbals, Their Origin and Evolution,* 2nd edition. Cambridge: Cambridge University Press.

Chadwick, J., and J. Marsh, eds. 1990. *Bioactive Compounds from Plants.* Ciba Foundation Symposium 185. Chichester, England: John Wiley & Sons.

Connolly, B., and R. Anderson. 1988. *First Contact.* New York: Penguin.

Harshberger, J. W. 1896. The purposes of ethno-botany. *The American Antiquarian* 17(2):73–81.

Kanny Lall Dev, R. B. 1896. *The Indigenous Drugs of India.* Calcutta, India: Thacker & Spink.

Mabberly, D. J. 1987. *The Plant Book: A Portable Dictionary of the Higher Plants.* Cambridge: Cambridge University Press.

Spradley, J. P. 1980. *Participant Observation.* New York: Holt, Rinehart, and Winston.

Chapter 2: Plants That Heal

Arvigo, R., and M. Balick. 1993. *Rainforest Remedies: One Hundred Healing Herbs of Belize.* Twin Lakes, WI: Lotus Press.

Balick, M. J., E. Elisabetsky, and S. Laird, eds. 1995. *Medicinal Resources of the Tropical Forest: Biodiversity and Its Importance to Human Health.* New York: Columbia University Press.

Boyd, M. R., and K. D. Paull. 1995. Some practical considerations and applications of the National Cancer Institute in vitro anticancer drug discovery screen. *Drug Development Research* 34:91–109.

Brooker, S. G., R. C. Cambie, and R. C. Cooper. 1987. *New Zealand Medicinal Plants.* Auckland, New Zealand: Reed Books.

Bye, R. A., Jr. 1986. Voucher specimens in ethnobiological studies and publications. *Journal of Ethnobiology* 6(1):1–8.

Chadwick, D. J., and J. Marsh, eds. 1994. *Ethnobotany and the Search for New Drugs.* Ciba Foundation Symposium 185. Chichester, England: John Wiley & Sons.

Cox, P. A., and M. J. Balick. 1994. The ethnobotanical approach to drug discovery. *Scientific American* 270(6):82–87.

Cox, P. A., L. R. Sperry, M. Tuominen, and L. Bohlin. 1989. Pharmacological activity of the Samoan ethnopharmacopoeia. *Economic Botany* 43:487–497.

Davis, E. W., and J. A. Yost. 1983. The ethnomedicine of the Waorani of Amazonian Ecuador. *Journal of Ethnopharmacology* 9:273–297.

De Smet, P. A. G. M. 1991. Is there any danger in using traditional medicine? *Journal of Ethnopharmacology* 31:181–192.

Duke, J. A. 1985. *CRC Handbook of Medicinal Herbs.* Boca Raton, FL: CRC Press.

Elisabetsky, E. 1986. New directions in ethnopharmacology. *Journal of Ethnobiology* 6(1):121-128.

Etkin, N. L. 1993. Anthropological methods in ethnopharmacology. *Journal of Ethnopharmacology* 38:93–104.

Evans, F. J. 1986. *Naturally Occurring Phorbol Esters.* Boca Raton, FL: CRC Press.

Farnsworth, N. R., and D. D. Soejarto. 1991. Global importance of medicinal plants. In O. Akerele, V. Heywood, and H. Synge, eds. *Conservation of Medicinal Plants.* Cambridge: Cambridge University Press.

Gerson, S. 1993. *Ayurveda: Ancient Indian Healing Art.* Shaftsburg, England: Element.

Gilman, A. G., L. S. Goodman, and A. Gilman, eds. 1980. *The Pharmacological Basis of Therapeutics,* 6th edition. New York: Macmillan.

Griggs, B. 1991. *Green Pharmacy.* Rochester, VT: Healing Arts Press.

Gustafson, K. R., J. H. Cardellina, J. B. McMahon, et al. 1992. A non-promoting phorbol from the Samoan medicinal plant *Homalanthus nutans* inhibits cell killing by HIV-1. *Journal of Medicinal Chemistry* 35:1978–1986.

Hartwell, J. L. 1982. *Plants Used Against Cancer.* Lawrence, MA: Quarterman Publications.

Hodge, W. H. 1948. Wartime cinchona procurement in Latin America. *Economic Botany* 2(3):229–257.

Holmstedt, B., S. H. Wassén, and R. E. Schultes. 1979. Jaborandi: An interdisciplinary appraisal. *Journal of Ethnopharmacology* 1:3–21.

Hostettmann, K., A. Marston, M. Maillard, and M. Hamburger. 1995. *Phytochemistry of Plants Used in Traditional Medicine.* Oxford: Clarendon Press.

Iwu, M. M. 1993. *Handbook of African Medicinal Plants.* Boca Raton, FL: CRC Press.

Jaramillo-Arango, J. 1950. *The Conquest of Malaria.* London: William Heinemann Medical Books Ltd.

King, S. R., and T. J. Carlson. 1995. Biocultural diversity, biomedicine and ethnobotany: The experience of Shaman Pharmaceuticals. *Interciencia* 20(3):134–139.

Kreig, M. B. 1964. *Green Medicine.* Chicago: Rand McNally.

Lewis, W. H., and M. P. F. Elvin-Lewis. 1977. *Medical Botany: Plants Affecting Man's Health.* New York: John Wiley & Sons.

———. 1995. Medicinal plants as sources of new therapeutics. *Annals of the Missouri Botanical Garden* 82:16–24.

Li, D., N. L. Owen, P. Perera, et al. 1994. Structure elucidation of three triterpenoid saponins from *Alphitonia zizyphoides* using 2D NMR techniques. *Journal of Natural Products* 57(2):218–224.

Moerman, D. E. 1986. *Medicinal Plants of Native America,* Vols. I & II. Technical Reports, Number 19. Ann Arbor: Museum of Anthropology, University of Michigan.

Nigg, H. N., and D. Seigler, eds. 1992. *Phytochemical Resources for Medicine and Agriculture.* New York: Plenum Press.

Oliver-Bever, B. 1986. *Medicinal Plants in Tropical West Africa.* Cambridge: Cambridge University Press.

Plotkin, M. J. 1993. *Tales of a Shaman's Apprentice: An Ethnobotanist Searches for New Medicines in the Amazon Rain Forest.* New York: Viking Penguin.

Samuelsson, G. 1992. *Drugs of Natural Origin.* Stockholm: Swedish Pharmaceutical Press.

Stetter, C. 1993. *The Secret Medicine of the Pharaohs: Ancient Egyptian Healing.* Chicago: Edition Q.

Tyler, V. E., L. R. Brady, and J. E. Robbers. 1988. *Pharmacognosy.* Philadelphia: Lea & Febiger.

Ubillas, R., S. D. Jolad, R. C. Bruening, et al. 1995. SP-303, an antiviral oligomeric proanthocyanidin from the latex of *Croton lechleri* (Sangre de Drago). *Phytomedicine* 1:77–106.

Vogel, V. J. 1970. *American Indian Medicine.* Norman, OK: University of Oklahoma Press.

Wagner, H., and N. R. Farnsworth, eds. 1990. *Plants and Traditional Medicine.* Economic & Medicinal Plant Research, Vol. 4. London: Academic Press.

Whistler, A. W. 1992. *Polynesian Herbal Medicine.* Lawai, HI: National Tropical Botanical Garden.

Chapter 3: From Hunting and Gathering to Haute Cuisine

Bennett, P. H., C. Bogardus, J. Tuomilehto, and P. Zimmet. 1992. Epidemiology and natural history of NIDDM: Non-obese and obese. *International Textbook of Diabetes Mellitus,* 147–176.

Bisset, N. G. 1989. Arrow and dart poisons. *Journal of Ethnopharmacology* 25:1–41.

Crosby, A. W. 1972. *The Columbian Exchange: Biological and Cultural Consequences of 1492.* Westport, CT: Greenwood Press.

———. 1986. *Ecological Imperialism: The Biological Expansion of Europe 900–1900.* Cambridge: Cambridge University Press.

Etkin, N., ed. 1994. *Eating on the Wild Side.* Tucson: University of Arizona Press.

Etkin, N., and P. Ross. 1982. Food as medicine and medicine as food: An adaptive framework for the interpretation of plant utilization among the Hausa of northern Nigeria. *Social Science and Medicine* 16:1559–1573.

———. 1991. Should we set a place for diet in ethnopharmacology? *Journal of Ethnopharmacology* 32:25–36.

Ford, R. I., ed. *Prehistoric Food Production in North America.* Ann Arbor: Museum of Anthropology, University of Michigan.

Fowler, C. and P. Mooney. 1990. *Shattering: Food, Politics, and the Loss of Genetic Diversity.* Tucson: University of Arizona Press.

Harlan, J. R. 1975. *Crops and Man.* Madison, WI: American Society of Agronomy.

Harris, D. R., and C. G. Hillman, eds. 1989. *Foraging and Farming: The Evolution of Plant Exploitation.* London: Unwin Hyman.

Heiser, C. B., Jr. 1990. *Seed to Civilization — The Story of Food.* Cambridge, MA: Harvard University Press.

Johns, T. 1990. *With Bitter Herbs They Shall Eat It: The Pharmacologic, Ecologic, and Social Implications of Using Noncultigens.* Tucson: University of Arizona Press.

———, E. B. Mhoro, P. Sanaya, and E. K. Kimanani. 1994. Herbal remedies of the Batemi of Ngorongoro District, Tanzania: A quantitative appraisal. *Economic Botany* 48:90–95.

Kirch, P. V. 1984. *The Evolution of the Polynesian Chiefdoms.* Cambridge: Cambridge University Press.

Nabhan, G. P. 1989. *Enduring Seeds.* San Francisco: North Point Press.

Sauer, J. D. 1994. *Historical Geography of Crop Plants.* Boca Raton, FL: CRC Press.

Thorburn, A.W., J. C. Brand, and A. S. Truswell. 1987. Slowly digested and absorbed carbohydrates in traditional bushfoods: A protective factor against diabetes. *American Journal of Clinical Nutrition* 45:98–106.

Vioa, H. J., and C. Margolis. 1991. *Seeds of Change.* Washington, D.C.: Smithsonian Institution Press.

Walston, J., C. Bogardus, K. Silver, et al. 1995. Time of onset of non-insulin-dependent diabetes mellitus and genetic variation in the β-adrenergic-receptor gene. *New England Journal of Medicine* 333:343–347.

Watson, D. 1961. *Indians of Mesa Verde.* Mesa Verde National Park, CO: Mesa Verde Museum Association.

Chapter 4: Plants as the Basis for Material Culture

Balée, W. 1994. *Footprints of the Forest: Ka'apor Ethnobotany—the Historical Ecology of Plant Utilization by an Amazonian People.* New York: Columbia University Press.

Banack, S. A., and P. A. Cox. 1987. Ethnobotany of ocean-going canoes in Lau, Fiji. *Economic Botany* 41:148–162.

Berlin, B., D. E. Breedlove, and P. H. Raven. 1973. General principles of classification and nomenclature in folk botany. *American Anthropologist* 75:214–242.

Best, E. 1977. *Forest Lore of the Maori.* Wellington, New Zealand: Government Printer.

Bisset, N. G. 1991. One man's poison, another man's medicine? *Journal of Ethnopharmacology* 32:71–81.

Brooker, S. G., R. C. Cambie, and R. C. Cooper. 1988. *Economic Native Plants of New Zealand.* Christchurch, New Zealand: Division of Industrial and Scientific Research.

Brown, C. H. 1985. Mode of subsistence and folk biological taxonomy. *Current Anthropology* 26:43–64.

Cannon, J., and M. Cannon. 1994. *Dye Plants and Dyeing.* Portland, OR: Timber Press.

Cox, P. A., and S. A. Banack, eds. 1991. *Islands, Plants, and Polynesians.* Portland, OR: Dioscorides Press.

Densmore, F. 1928. *Uses of Plants by Chippewa Indians.* Washington, D.C.: Government Printing Office.

Finney, B. 1976. *Hokule'a: The Way to Tahiti.* New York: Dodd, Mead, and Co.

Gils, C. G., and P. A. Cox. 1994. Ethnobotany of nutmeg in the Spice Islands. *Journal of Ethnopharmacology* 42:117–124.

Greenberg, S., and E. L. Ortiz. *The Spice of Life.* New York: The Amaryllis Press.

Hanna, W. A. 1978. *Indonesian Banda.* Philadelphia: The Institute for the Study of Human Issues.

Hardy, D. E. 1991. *Tatoo Time: Art from the Heart.* Honolulu: Hardy Marks Publication.

Heyerdahl, T. 1989. *Easter Island—The Mystery Solved.* Toronto: Soddart Publishing.

Hiroa, T. R. (Sir Peter Buck). 1982. *The Coming of the Maori.* Wellington, New Zealand: Whitcoulls.

Jennings, J. D., ed. *The Prehistory of Polynesia.* Cambridge, MA: Harvard University Press.

Kocher Schmid, C. 1991. *Of People and Plants: A Botanical Ethnography of Nokop Village, Madang and Morobe Provinces, Papua New Guinea.* Ethnologisches Seminar der Universität und Museum für Völkerkunde. Basel, Switzerland: Wepf & Co. A.G. Verlag.

Levinson, M., R. G. Ward, and J. W. Webb. 1973. *The Settlement of Polynesia: A Computer Simulation.* Minneapolis, MN: University of Minnesota Press.

Lewis, D. 1972. *We, the Navigators.* Honolulu: University of Hawaii Press.

McClatchey, W., and P. A. Cox. 1992. Use of the Sago palm *Metroxylon warburgii* in the Polynesian island Rotuma. *Economic Botany* 46:305–309.

Norman, J. 1990. *The Complete Book of Spices.* New York: Penguin.

Purseglove, J. W., E. G. Brown, C. L. Green, and S. R. J. Robbins. 1981. *Spices,* Vols. I & II. London: Longman.

Sauer, J. D. 1988. *Plant Migration: The Dynamics of Geographical Patterning in Seed Plant Species.* Berkeley: University of California Press.

Sneider, C. and W. Kyselka. 1986. *The Wayfinding Art: Ocean Voyaging in Ancient Polynesia.* Berkeley: Regents of the University of California.

Yen, D. E. 1974. *The Sweet Potato and Oceania: An Essay in Ethnobotany.* Bernice P. Bishop Museum Bulletin 236. Honolulu: Bishop Museum Press.

Chapter 5: Entering the Other World

Andrews, G., and D. Solomon, eds. 1975. *The Coca Leaf and Cocaine Papers.* New York: Harcourt Brace Jovanovich.

Arvigo, R., N. Epstein, and M. Yaquinto. 1994. *Sastun: My Apprenticeship with a Maya Healer.* San Francisco: Harper San Francisco.

Carter, Steven. 1993. *The Culture of Disbelief: How American Law and Politics Trivialize Religious Devotion.* New York: Basic Books.

Davis, W. 1992. *Shadows in the Sun: Essays on the Spirit of Place.* Alberta, Canada: Lone Pine Publishing.

Drury, N. 1989. *The Elements of Shamanism.* Shaftsbury, England: Element.

Efron, D. H., B. Holmstedt, and N. S. Kline. 1979. *Ethnopharmacologic Search for Psychoactive Drugs.* New York: Raven Press.

Emboden, W. 1979. *Narcotic Plants: Hallucinogens, Stimulants, Inebriants and Hypnotics, Their Origins and Uses.* New York: Collier Books.

Furst, P. T., ed. 1972. *Flesh of the Gods: The Ritual Use of Hallucinogens.* New York: Praeger Publishers.

Golden, M. W. 1974. *History of Coca: The "Divine Plant" of the Incas.* Fitz Hugh Ludlow Memorial Library Edition. San Francisco: And/or Press.

Harner, M. J., ed. 1973. *Hallucinogens and Shamanism.* London: Oxford University Press.

Herer, J. 1990. The forgotten history of hemp. *Earth Island Journal* (Fall):35–39.

Hoffman, A. 1983. *LSD: My Problem Child.* New York: Putnam Publishing Group.

Holmstedt, B. 1972. The ordeal bean of old Calabar: The pagent of *Physostigma venenosum* in medicine. In T. Swain, ed. *Plants in the Development of Modern Medicine,* pp. 303–360. Cambridge, MA: Harvard University Press.

Lebot, V., M. Merlin, and L. Lindstrom. 1992. *Kava, the Pacific Drug.* New Haven, CT: Yale University Press.

Lebot, V., and P. Cabalion. 1986. *Les Kavas de Vanuatu.* Paris: Éditions de l'ORSTROM.

Lewin, L. 1931. *Phantastica: Narcotic and Stimulating Drugs — Their Use and Abuse.* New York: E. P. Dutton.

Luna, L. E., and P. Amaringo. 1991 *Ayahuasca Visions: The Religious Iconography of a Peruvian Shaman.* Berkeley, CA: North Atlantic Books.

Plowman, T. 1984. The origin, evolution, and diffusion of coca, *Erythroxylum* spp., in South and Central America. In D. Stone, ed. *Pre-Columbian Plant Migration, Papers of the Peabody Museum of Archaeology and Ethnology* 76:129–163.

———. 1981. Amazonian coca. *Journal of Ethnopharmacology* 3:195–225.

———. 1984. The ethnobotany of coca (*Erythroxylum* spp., Erythroxylaceae). *Advances in Economic Botany* 1:62–111.

Schultes, R. E., and A. Hofmann. 1979. *Plants of the Gods: Origins of Hallucinogenic Use.* New York: McGraw-Hill.

———. 1980. *The Botany and Chemistry of Hallucinogens,* 2nd edition. Springfield, IL: Charles C. Thomas.

Stone, T. W. 1995. *Neuropharmacology.* New York: W. H. Freeman.

Tullis, F. LaMond. 1995. *Unintended Consequences: Illegal Drugs and Drug Policies in Nine Countries.* Boulder, CO: Lynne Rienner Publishers.

Weil, A. 1995. Letter from the Andes: The new politics of coca. *The New Yorker* (May 15):70–80.

Zethelius, M., and M. J. Balick. 1982. Modern medicine and shamanistic ritual: A case of positive synergistic response in the treatment of a snakebite. *Journal of Ethnopharmacology* 5(2):181–185.

Zias, J., H. Stark, J. Seligman, et al. 1993. Early medical use of cannabis. *Nature* 363:215.

Chapter 6: Biological Conservation and Ethnobotany

Alcorn, J. B. 1984. *Huastec Mayan Ethnobotany.* Austin: University of Texas Press.

Anderson, A. B., P. H. May, and M. J. Balick. 1991. *The Subsidy from Nature: Palm Forests, Peasantry and Development on an Amazonian Frontier.* New York: Columbia University Press.

Anderson, A. B., and E. M. Ioris. 1992. Valuing the rain forest: Economic strategies by small-scale forest extractivists in the Amazon Estuary. *Human Ecology* 20(3):337–369.

Balick, M. J., and R. Mendelsohn. 1992. Assessing the economic value of traditional medicines from tropical rain forests. *Conservation Biology* 6(1):128–130.

Berkes, F., D. Feeny, B. J. McCay, and J. M. Acheson. 1989. The benefits of the commons. *Nature* 340:91–93.

Boom, B. M. 1985. "Advocacy Botany" for the Neotropics. *Garden* May/June 1985:24–32.

Bye, R. A., Jr. 1986. Medicinal plants of the Sierra Madre: Comparative study of Tarahumara and Mexican market plants. *Economic Botany* 40(1):103–124.

Caballero, J. 1994. Use and Management of Sabal Palms Among the Maya of Yucatan. Ph.D. dissertation, Department of Anthropology, University of California at Berkeley.

Cox, P. A., and T. Elmqvist. 1991. Indigenous control of tropical rainforest reserves: An alternative strategy for conservation. *Ambio* 20(7):317–321.

———. 1993. Ecocolonialism and indigenous knowledge systems: Village controlled rainforest preserves in Samoa. *Pacific Conservation Biology* 1:11–25.

Greaves, T. 1994. *Intellectual Property Rights for Indigenous Peoples: A Source Book.* Oklahoma City: Society for Applied Anthropology.

Grimes, A., S. Loomis, P. Jahnige, et al. 1994. Valuing the rain forest: The economic value of nontimber forest products in Ecuador. *Ambio* 23(7):405–410.

Hardin, G. 1968. The tragedy of the commons. *Science* 162:1243–1248.

Kemf, E. 1993. *The Law of the Mother: Protecting People in Protected Areas.* San Francisco: Sierra Club Books.

King, S. R. 1994. Establishing reciprocity: Biodiversity, conservation and new models for cooperation between forest-dwelling peoples and the pharmaceutical industry. In T. Greaves, ed. *Intellectual Property Rights for Indigenous Peoples, A Sourcebook,* pp. 69–82. Oklahoma City: The Society for Applied Anthropology.

MacKenzie, J. M. 1990. *Imperialism and the Natural World.* Manchester, England: Manchester University Press.

May, P. H., A. B. Anderson, M. J. Balick, and J. M. F. Frazão. 1985. Babassu palm agroforestry systems in Brazil's Mid-North Region. *Agroforestry Systems* 3:275–295.

Peters, C. M., A. H. Gentry, and R. O. Mendelsohn. 1989. Valuation of an Amazonian rainforest. *Nature* 339:655–656.

Pinedo-Vasquez, M., D. Zarin, P. Jipp, and J. Chota-Inuma. 1990. Use-values of tree species in a communal forest reserve in Northeast Peru. *Conservation Biology* 4(4):405–416.

Plotkin, M., and L. Famolare, eds. 1992. *Sustainable Harvest and Marketing of Rain Forest Products.* Washington, D.C.: Island Press.

Prance, G. T., W. Balée, B. M. Boom, and R. L. Carneiro. 1987. Quantitative ethnobotany and the case for conservation in Amazonia. *Conservation Biology* 1:296–310 (December).

Additional information on indigenous–controlled reserves in Samoa can be obtained from the Seacology Foundation, P.O. Box 400, Springville, Utah 84663 or on the Internet at WWW.Seacology.org.

Sources of Illustrations

Emil Huston rendered the illustrations on pages 12, 79, 152–153, 164, and 172. Dolores R. Santoliquido rendered the illustrations on pages 73 and 190. The remaining illustrations were rendered by Fine Line Illustrations.

Chapter 1

Facing page 1: Detail of the mural, "History of Medicine." Diego Rivera. The Hospital de la Raza, Mexico City. Schalkwijk/Art Resource, NY

Page 2: Michael Balick

Page 3: University of Pennsylvania Archives

Page 4: John Wang, NuSkin International, Inc.

Page 6: Michael Balick

Page 8: Mark Dell'Aquila

Page 10: Michael Balick

Page 11: Elysa Hammond

Page 14: Photographed by Mark Philbrick

Page 15: The Linnean Society of London

Page 16: "Portrait of William Withering." Carl F. von Breda. Nationalmuseum, SKM, Stockholm

Page 19: Nationaal Natuurhistorisch Museum, Leiden, The Netherlands

Page 20: Collection of Wade Davis

Page 21: Albert Hofmann

Page 22: Lynn Johnson/Black Star

Chapter 2

Page 24: Michael Balick

Page 27: Photographed by Michael Balick. The New York Botanical Garden

Page 30: José Cuatrecasas

Page 31: W. C. Steere, Courtesy of The New York Botanical Garden Libary.

Page 32: Photographed by Michael Balick. From J. Sturm. 1904. *Flora von Deutschland.* Stuttgart: Verlag von K. G. Lutz. The New York Botanical Garden

Page 36: Michael Balick

Page 37: Steven King

Page 38: From M. J. Balick. 1990. Ethnobotany and the identification of therapeutic agents from the rainforest. In D. J. Chadwick and J. Marsh (eds.). *Bioactive Compounds from Plants.* Ciba Foundation Symposium 154: 24–26. Chichester, England: John Wiley & Sons

Page 41: Michael Balick

Page 43: Mark Philbrick, Brigham Young University

Page 45: Michael Balick

Page 46: From M. J. Balick. 1990. Ethnobotany and the identification of therapeutic agents from the rainforest. In D. J. Chadwick and J. Marsh (eds.). *Bioactive Compounds from Plants.* Ciba Foundation Symposium 154: 30. Chichester, England: John Wiley & Sons

Page 47 *(left):* Michael Balick

Page 47 *(right):* Tori Butt

Page 48: The New York Botanical Garden

Page 54: Paul Cox

Pages 56 and 57: Modified and corrected from P. A. Cox. 1994. The ethnobotanical approach to drug discovery: Strengths and limitations. In G. Prance and J. Marsh (eds.). *Ethnobotany and the Search for New Drugs.* Ciba Foundation Symposium 185: 25–41. London: Academic Press

Page 58: Michael Balick

Page 59: Jean-Paul Ferrero/AUSCAPE

Chapter 3

Page 62: Michael Balick

Page 65: Blaine Furniss

Page 69: Timothy Johns

Page 76: William H. Jackson, Museum of New Mexico

Page 77: Tom Till

Page 79: Drawings based on photographs in Robert H. Lister and Florence C. Lister. 1978. *Anasazi Pottery.* Albuquerque: Maxwell Museum of Anthropology and the University of New Mexico Press

Page 82: Bishop Museum

Page 84: Bishop Museum

Page 85: Paul Cox

Page 86: From P. A. Cox. 1994. Wild plants as food and medicine in Polynesia. In N. L. Etkin (ed.), pp. 85–113. *Eating on the Wild Side.* Tucson: University of Arizona Press

Page 88: Paul Cox

Page 92: Steven King

Page 94: Robert C. Clarke

Page 96: Royal Botanic Gardens, Kew

Chapter 4

Page 98: "Tahiti, Bearing South East." William Hodges, 1773. National Maritime Museum, London

Page 104: Sandra Banack

Page 105: Sandra Banack

Page 106: Modified from S. A. Banack and P. A. Cox. 1987. Ethnobotany of ocean-going canoes in Lau, Fiji. *Economic Botany* 41: 148–162

Page 107: Sandra Banack

Page 108: LaMoyne Garside

Page 111: Don Haddon/Ardea, London

Page 112: Data from D. E Yen. 1974. The sweet potato and Oceania: An essay in ethnobotany. *Bernice P. Bishop Museum Bulletin* 236

Page 113: Paul Cox

Page 114: The Museum of New Zealand. Te Papa Tongarewa, Wellington, New Zealand

Page 115: Modified from N. G. Bisset. 1989. Arrow and dart poisons. *Journal of Ethnopharmacology* 25: 1–41

Page 117: John Wright/Hutchison Library

Page 118: Based on data from N. G. Bisset. 1989. Arrow and dart poisons. *Journal of Ethnopharmacology* 25: 1–41

Page 121 *(left):* Mark Dell'Aquila

Page 121 *(right):* Art Whistler

Page 122 *(left):* Bishop Museum

Page 122 *(right):* Mark Dell'Aquila

Page 124: Anthony Anderson

Page 126: Paul Cox

Page 127: Paul Cox

Page 128: E. Best. 1942. *Forest Lore of the Maori,* 101. Wellington, New Zealand: Goverment Printer

Page 129: Paul Cox

Page 130: Photographed by Paul Cox. From T. R. Peale. 1848. *Mammalia and Ornithology in U. S. Exploring Expedition 1838–1842,* Vol. 8. Philadelphia: United States Government

Page 131: Paul Cox

Page 133: Paul Cox

Page 135: Paul Cox

Page 138: Paul Cox

Page 141: Mark Philbrick, Brigham Young University

Chapter 5

Page 144: "The Induction of Aya-huasca in the Brain." Pablo Amaringo, 1995. Artist Management International

Page 147: Kohl, Friedrich Georg, 1895. Die officinellen Pflanzen der Pharmacopoea Germanica fur Pharmeceuten und Mediciner, J. A. Barth. Leipzig. Courtesy of The New York Botanical Garden Library

Page 149: Arnold Newman/Peter Arnold, Inc.

Page 155: Michael Balick

Pages 156–157: Redrawn based on maps published in Richard Evans Schultes and Albert Hoffman. 1992. *Plants of the Gods: Their Sacred, Healing, and Hallucinogenic Powers.* Rochester, VT: Healing Arts Press

Page 161: Jean-Paul Ferrero/AUSCAPE

Page 163: Bishop Museum

Page 164: Redrawn based on a drawing in M. I. Artamonov. May 1965. Frozen Tombs of the Scythians. *Scientific American* 212, No. 5: 108

Page 166: Steven King

Page 167: Collection of Wade Davis

Page 168: Collection of Wade Davis

Page 170: Steven King

Page 172: Drawing based on photograph published in Mark D. Merlin, fig. 83. 1984. *On the Trail of the Ancient Opium Poppy.* Cranbury, N. J.: Associated University Presses

Page 173: Scott Carnazine/Photo Researchers

Page 174: From *Curtis's Botanical Magazine* III. 1847. Table 4296. Courtesy of The New York Botanical Garden Library

Page 175: Tom Carlson

Chapter 6

Page 178: Michael Balick

Page 180: Thomas Elmqvist

Page 181: John Wang, NuSkin International, Inc.

Page 184: Michael Balick

Page 185: Nature 339 (25 June 1989): 655.

Page 187: Michael Balick

Page 188: Macduff Everton

Page 190: From A. B. Anderson. 1988. Use and management of native forests dominated by Açaí palm (Euterpe oleracea Mart.) in the Amazon Estuary. "The Palm—Tree of Life," *Advances in Economic Botany* 6: 149. Copyright 1988, The New York Botanical Garden

Page 193: Jean-Paul Ferrero/AUSCAPE

Page 194: Michael Balick

Page 196: Michael Balick

Page 202: Michael Balick

Page 204: Paul Cox

Index

Note: Page numbers in *italics* indicate illustrations; t indicates tables.